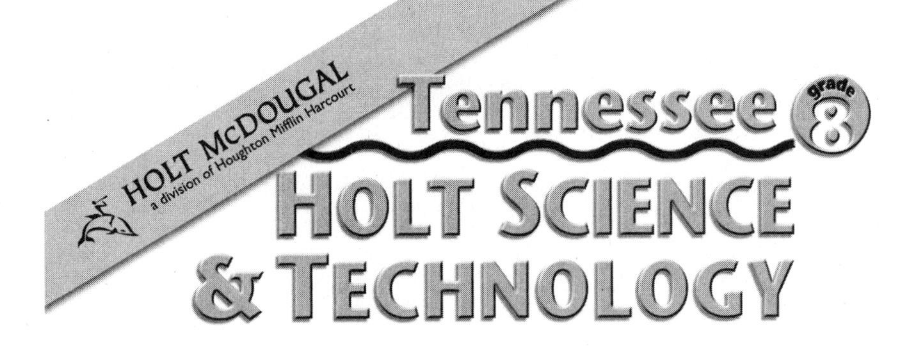


# Reading Comprehension Guide

Printed in the United States of America



ISBN 13: 978-0-55-401779-2
ISBN 10: 0-55-401779-2

1 2 3 4 5 6 054 12 11 10 09 08

# Contents

# Contents

Name______________________________ Class ________________ Date ________________

Skills Worksheet

# Directed Reading A

## Section: Science and Scientists

1. What are the first two steps in being a scientist?

______________________________________________________________________

______________________________________________________________________

### SCIENCE STARTS WITH A QUESTION

2. What is science?

______________________________________________________________________

______________________________________________________________________

3. Explain how you might practice science in your own neighborhood.

______________________________________________________________________

______________________________________________________________________

4. What are three different kinds of environments you might ask questions about?

______________________________________________________________________

______________________________________________________________________

______________________________________________________________________

### INVESTIGATION: THE SEARCH FOR ANSWERS

**Match the correct definition with the correct term. Write the letter in the space provided.**

_____ 5. carefully looking and recording what you see

_____ 6. performing an activity to answer questions

_____ 7. looking up information in books or the Internet

a. research
b. experimentation
c. observation

### WHY ASK WHY?

8. What are two ways science has made automobiles safer?

______________________________________________________________________

______________________________________________________________________

9. What are three natural resources that are saved by recycling steel?

______________________________________________________________

______________________________________________________________

______________________________________________________________

10. How have chlorofluorocarbons harmed the environment?

______________________________________________________________

______________________________________________________________

11. What are the results of damaging the ozone layer?

______________________________________________________________

______________________________________________________________

## SCIENTISTS ARE ALL AROUND YOU

**Match the correct definition with the correct term. Write the letter in the space provided.**

_____ 12. a person who studies a community of organisms and their environment

_____ 13. a person who draws scientific diagrams

_____ 14. a scientist who studies the chemistry of rocks, minerals, and soil

_____ 15. a person who studies the atmosphere

_____ 16. a scientist who studies volcanoes

a. meteorologist
b. volcanologist
c. science illustrator
d. ecologist
e. geochemist

17. What are two careers that a meteorologist might have?

______________________________________________________________

______________________________________________________________

18. What are two reasons why a geochemist might study rocks?

______________________________________________________________

______________________________________________________________

19. What are four fields an ecologist might work in?

_______________________________________________________________

_______________________________________________________________

_______________________________________________________________

_______________________________________________________________

20. How can a volcanologist help save lives?

_______________________________________________________________

_______________________________________________________________

21. What two subjects do most science illustrators have a background in?

_______________________________________________________________

_______________________________________________________________

Name______________________________ Class ________________ Date ________________

Skills Worksheet

# Directed Reading A

## Section: Scientific Methods

1. What did James Czarnowski and Michael Triantafyllou develop?

_______________________________________________________________

### WHAT ARE SCIENTIFIC METHODS?

2. The ways scientists answer questions and solve problems are called

_______________________.

3. Do scientists always use the steps in the scientific method in the same way? Explain your answer.

_______________________________________________________________

_______________________________________________________________

_______________________________________________________________

### ASK A QUESTION

**Write the letter of the correct answer in the space provided.**

_____ 4. What step helps scientists focus the purpose of an investigation?
 a. using the scientific methods
 b. asking a question
 c. using observation
 d. finding a dinosaur bone

_____ 5. Which statement about observations is true?
 a. They always involve seeing something.
 b. They must be made at the beginning of an investigation.
 c. They involve noticing the natural world.
 d. They can be made with any of the senses.

_____ 6. Why is it important for observations to be accurately recorded?
 a. so that scientists will be paid
 b. so that scientists never need to make the same observations
 c. so that other scientists will not be able to use the information
 d. so that scientists can use the information in future investigations

_____ 7. What is a measurement?
 a. A measurement is a tool.
 b. A measurement is a recording.
 c. A measurement is an observation.
 d. A measurement is a way to ask a question.

**Match the correct definition with the correct term. Write the letter in the space provided.**

_____ 8. a comparison of energy output with energy input

_____ 9. application of science for practical purposes

_____ 10. scientist who puts scientific knowledge to practical use

a. technology
b. engineer
c. efficiency

11. What did Czarnowski and Triantafyllou learn from observing boat propulsion systems?

________________________________________________________________

12. Explain why a boat's propeller system is not as efficient as it could be.

________________________________________________________________

13. What is the formula used to calculate efficiency?

________________________________________________________________

14. What question did Czarnowski and Triantafyllou ask?

________________________________________________________________

## FORM A HYPOTHESIS

**Write the letter of the correct answer in the space provided.**

_____ 15. A possible explanation or answer to a question is a(n)
a. prediction.
b. data.
c. variable.
d. hypothesis.

_____ 16. A good hypothesis
a. is always true.
b. always answers a scientific question.
c. is testable.
d. includes a lot of information.

_____ 17. What hypothesis did Czarnowski's observations lead to?
a. The penguin propulsion system is more efficient than propellers.
b. The penguin propulsion system is less efficient than propellers.
c. Propulsion systems cannot be improved.
d. Penguins have rigid bodies.

_____ 18. A statement in an if-then format is a(n)
a. prediction.
b. data.
c. variable.
d. hypothesis.

**TEST THE HYPOTHESIS**

_____ 19. How would you find out if your hypothesis is a reasonable answer to your question?
a. Make more observations.
b. Look up the answer in an encyclopedia.
c. Test the hypothesis.
d. Develop a new hypothesis.

20. A test that compares the results from a control group with the results from one or more experimental groups is a(n) ______________________.

21. The one factor that changes in a controlled experiment is called a(n) ______________________.

22. How did Czarnowski and Triantafyllou test their hypothesis?

______________________________________________________________

23. Pieces of information gathered through observation or experimentation are called ______________________.

24. Why do scientists like to study a large number of samples?

______________________________________________________________

______________________________________________________________

______________________________________________________________

______________________________________________________________

25. What were three kinds of data the scientists collected about *Proteus*?

______________________________________________________________

______________________________________________________________

______________________________________________________________

## ANALYZE THE RESULTS

**Write the letter of the correct answer in the space provided.**

_____ 26. What must scientists do after they have collected their data?
a. Make a prediction.
b. Analyze results.
c. Communicate results.
d. Draw a conclusion.

_____ 27. How was the data collected by Czarnowski and Triantafyllou organized?
a. in a speech
b. as photographs
c. on graphs
d. as an experiment

## DRAW CONCLUSIONS

_____ 28. Which is NOT a conclusion a scientist might draw at the end of an investigation?
a. The investigation was a waste.
b. The results support the hypothesis.
c. The hypothesis was not proven.
d. More information is needed.

29. What hypothesis was supported by Czarnowski and Triantafyllou's investigation?

_______________________________________________________________

_______________________________________________________________

## COMMUNICATE RESULTS

_____ 30. Why do scientists communicate the results of their investigations?
a. so no other scientist will need to make the same experiment
b. so other scientists can reproduce the experiment and verify the data
c. so credit will be given to the right scientist
d. to prove the hypothesis is true

31. What were three places the *Proteus* results were shared?

_______________________________________________________________

_______________________________________________________________

_______________________________________________________________

Name______________________________ Class ________________ Date ________________

Skills Worksheet

# Directed Reading A

## Section: Scientific Models

### TYPES OF SCIENTIFIC MODELS

_____ 1. A representation of an object or system is a(n)
a. hypothesis.
b. mimic.
c. model.
d. experiment.

_____ 2. How does a model help people understand the natural world?
a. Models use familiar objects or ideas that stand for other things.
b. Models use unfamiliar objects or ideas that stand for other things.
c. Models have no limitations to showing the structure of an object.
d. Models always look exactly like the objects they represent.

_____ 3. A model that looks like the thing it represents is a
a. physical model.
b. mathematical model.
c. conceptual model.
d. hypothetical model.

_____ 4. A model that is made up of mathematical equations and data is a
a. physical model.
b. mathematical model.
c. conceptual model.
d. hypothetical model.

_____ 5. A system of ideas or a comparison with familiar things is a
a. physical model.
b. mathematical model.
c. conceptual model.
d. hypothetical model.

6. What is one example of a limitation of a physical model?

______________________________________________________________

7. What is one example of a limitation of a mathematical model?

______________________________________________________________

8. What is one example of a limitation of a conceptual model?

______________________________________________________________

## MODELS ARE JUST THE RIGHT SIZE

9. What is one example of a large thing that can be represented by a model?

_______________________________________________

10. What is one example of a small thing that can be represented by a model?

_______________________________________________

## MODELS BUILD SCIENTIFIC KNOWLEDGE

**Write the letter of the correct answer in the space provided.**

_____ 11. An explanation for an event based on observation, experimentation, and reasoning is a(n)
a. law.
b. hypothesis.
c. variable.
d. theory.

_____ 12. Which statement about a theory is NOT true?
a. It can predict what might happen in the future.
b. It may be illustrated by a model.
c. It may be based on a guess.
d. It has been supported by experiments.

_____ 13. What can happen to a model when scientists change their theories because of new observations?
a. The model always stays the same.
b. The model will be eliminated.
c. The model may be changed or replaced.
d. Different scientists must make a new model.

_____ 14. A summary of many experimental results and observations is a(n)
a. law.
b. hypothesis.
c. variable.
d. theory.

_____ 15. What is true about a scientific law?
a. A law tells you how and why things work.
b. A law tells you what might happen.
c. A law tells you the same thing will happen most of the time.
d. A law tells you only what happens, not why it happens.

Name________________________________ Class __________________ Date __________________

Skills Worksheet

# Directed Reading A

## Section: Science and Engineering

_____ 1. What technology did scientists and engineers create to help you if your car breaks down?
a. jumper cables
b. telephone booth
c. cellular phone
d. computer

### WHAT IS TECHNOLOGY?

2. What is the definition of technology?

_______________________________________________

_______________________________________________

_______________________________________________

### HOW DOES SCIENCE RELATE TO TECHNOLOGY?

_____ 3. How does science relate to technology?
a. They are essentially the same.
b. Science does not require any engineering.
c. Engineering uses science to develop technology.
d. Science uses facts, engineering uses scientific laws.

_____ 4. Which of the following were developed by engineers?
a. biology, biotechnology, zoology
b. cellular phones, hybrid cars, disease-resistant corn
c. principles, theories, laws
d. physical, mathematical, conceptual

_____ 5. Besides new products, what else do engineers design?
a. tools and processes needed to make new products
b. designs for biological organisms
c. designs for new cellular telephones
d. theories that lead to more scientific investigations

_____ 6. What is the definition of engineering?
a. the products and processes designed to serve our needs
b. the application of science to help living organisms
c. the process of creating technology
d. finding ways to apply science to design

## WHAT IS THE ENGINEERING DESIGN PROCESS?

**Match the correct description with the correct step. Write the letter in the space provided.**

_____ 7. This is the first step of the engineering design process.

_____ 8. This step of the process is where engineers do a cost-benefit analysis.

_____ 9. Engineers might do some brainstorming in this step.

_____ 10. In this step, engineers create a test model.

_____ 11. Engineers might try new solutions in this step.

a. Making a Prototype
b. Modifying and Retesting the Solution
c. Testing and Evaluating
d. Identifying and Researching a Need
e. Developing Possible Solutions

## COMMUNICATION

12. Explain why it is important for engineers to communicate.

_______________________________________________________________

_______________________________________________________________

_______________________________________________________________

## TECHNOLOGY AND SOCIETY

_____ 13. Which of the following is an example of how engineers fulfill a social need?
a. providing information for police and firefighters
b. developing new materials for telephone and radio towers
c. writing software to improve collection of data from emergency calls
d. designing emergency call buttons for the elderly living alone

## INTENDED BENEFITS

_____ 14. Which of the following is an intended benefit of the cell phone?
a. cell phone relay towers
b. convenient communication
c. the creation of cell phone industry jobs
d. noise pollution

## UNINTENDED CONSEQUENCES

_____ 15. What is a positive unintended consequence of cell phone technology?
a. has provided many people with jobs
b. has made communication easier
c. has caused a lot of pollution
d. has resulted in relay towers that dominate the landscape

_____ 16. What is a negative unintended consequence of cell phone technology?
a. has made communication easier
b. has improved safety
c. has resulted in noise pollution
d. has provided many people with jobs

## BIOENGINEERING

17. What do bioengineers do?

_______________________________________________________________

_______________________________________________________________

_______________________________________________________________

## ASSISTIVE BIOENGINEERING

18. Give an example of assistive bioengineering.

_______________________________________________________________

_______________________________________________________________

_______________________________________________________________

## ADAPTIVE BIOENGINEERING

19. Give an example of adaptive bioengineering.

_______________________________________________________________

_______________________________________________________________

_______________________________________________________________

Skills Worksheet

# Directed Reading A

## Section: Tools, Measurement, and Safety

1. Anything that helps you do a task is a(n) ______________________.

### TOOLS FOR MEASURING

**Match the correct description with the correct term. Write the letter in the space provided.**

_____ 2. measures time

_____ 3. measures volume

_____ 4. measures temperature

_____ 5. measures length

_____ 6. measures force

_____ 7. measures mass

a. balance
b. thermometer
c. graduated cylinder
d. spring scale
e. meterstick
f. stopwatch

### TOOLS FOR ANALYZING

_____ 8. One way to find an average of data is by using
a. a pencil and graph paper.
b. a calculator.
c. a meterstick.
d. a stopwatch.

_____ 9. One way to create a graph to show data is by using
a. a pencil and graph paper.
b. a calculator.
c. a meterstick.
d. a stopwatch.

### MEASUREMENT

_____ 10. Some modern standardized units of measurement were originally based on
a. the weather.
b. mythology.
c. parts of the body.
d. ancient worldwide standards.

_____ 11. What was the International System of Units (SI) first called?
a. the French Academy of Sciences system
b. the cubic system
c. the International Measuring System
d. the metric system

_____ 12. What is one advantage of using the International System of Units?
a. Scientists can share and compare their observations and results.
b. Scientists can keep their observations and results to themselves.
c. All scientific results are accurate.
d. Inaccurate scientific results can be recorded.

_____ 13. Why is it an advantage that all SI units are based on the number 10?
a. Changing from one unit to another is not necessary.
b. All units can be measured in meters.
c. All measurements are accurate within .10 units.
d. Changing from one unit to another is easier.

_____ 14. What is the basic unit of length in the SI?
a. centimeter
b. kilometer
c. millimeter
d. meter

_____ 15. How many times larger than a decimeter is a meter?
a. 10
b. 100
c. 1000
d. .10

_____ 16. What measurement would a scientist use to measure the thickness of an ice sheet?
a. meter
b. cubic meter
c. inch
d. kilogram

_____ 17. A measure of the size of a surface or a region is the
a. density.
b. volume.
c. mass.
d. area.

_____ 18. What equation would be used to calculate the amount of wallpaper you would need for your classroom?
a. area = length + width
b. area = length × width
c. area = length × width × height
d. volume = length × width × height

_____ 19. A measure of the amount of matter in an object is its
a. density.
b. volume.
c. mass.
d. area.

_____ 20. What is the basic SI unit for mass?
a. kilogram (kg)
b. cubic meter ($m^3$)
c. kelvin (K)
d. nanometer (nm)

_____ 21. What SI unit would be best to use to measure the mass of a small object, like an apple?
a. kilogram (kg)
b. gram (g)
c. milligram (mg)
d. nanogram (ng)

_____ 22. The amount of space that something occupies, or that something contains, is called
a. density.
b. volume.
c. mass.
d. area.

_____ 23. What is the basic SI unit for volume?
a. kilogram (kg)
b. cubic meter ($m^3$)
c. kelvin (K)
d. nanometer (nm)

_____ 24. What tool is used to measure the volume of a liquid?
a. balance
b. spring scale
c. thermometer
d. graduated cylinder

_____ 25. What equation would be used to calculate the volume of a storage box?
a. volume = length + width + height
b. volume = length × width + height
c. volume = length × area
d. volume = length × width × height

_____ 26. The amount of matter in a given volume is its
a. density.
b. volume.
c. mass.
d. area.

_____ 27. What two measurements must be known in order to calculate density?
a. volume and area
b. mass and volume
c. mass and area
d. volume and cubic meters

_____ 28. What equation is used to calculate density?
a. density = mass ÷ volume
b. density = volume ÷ mass
c. density = volume + mass
d. density = volume × mass

29. The _______________ is a measure of how hot or cold something is.

30. The SI base unit for temperature is _______________.

31. Instead of using kelvins, scientists often measure temperature in a unit called

_______________.

**SAFETY RULES!**

**Match the correct description with the correct symbol. Write the letter in the space provided.**

_____ 32. sharp object

_____ 33. hand safety

_____ 34. chemical safety

_____ 35. eye protection

a.

b.

c.

d.

Name______________________________ Class ________________ Date ________________

Skills Worksheet

# Vocabulary and Section Summary

## Science and Scientists

### VOCABULARY

**In your own words, write a definition of the following term in the space provided.**

1. science

________________________________________________________________________

________________________________________________________________________

### SECTION SUMMARY

**Read the following section summary.**

- Three methods of investigation are research, observation, and experimentation.
- Science affects people's daily lives. Science can help save lives, save resources, and improve the environment.
- There are several types of scientists and many careers in science.
- Meteorologists study Earth's atmosphere.
- Geochemists study the chemistry of rocks, minerals, and soil.
- Ecologists study the behavior of living things.
- Volcanologists study volcanoes and Earth's structure and chemistry.

Name________________________________ Class ____________________ Date ____________________

Skills Worksheet

# Vocabulary and Section Summary

## Scientific Methods

### VOCABULARY

**In your own words, write a definition of the following terms in the space provided.**

1. scientific methods

_______________________________________________

_______________________________________________

2. observation

_______________________________________________

_______________________________________________

3. technology

_______________________________________________

_______________________________________________

4. hypothesis

_______________________________________________

_______________________________________________

5. data

_______________________________________________

_______________________________________________

## SECTION SUMMARY

**Read the following section summary.**

- Scientific methods are the ways in which scientists answer questions and solve problems.
- Asking a question helps you focus the purpose of an investigation.
- A hypothesis is a possible answer to a question. A good hypothesis is testable.
- Testing a hypothesis helps you find out if the hypothesis is a reasonable answer to your question.
- Analyzing the data collected during an investigation will help you find out whether the results of your test support your hypothesis.
- Conclusions that you draw from your results will show you if your test supported your hypothesis.
- Communicating your results will allow other scientists to use your investigation for research or conduct an investigation of their own.

Name_______________________________ Class ___________________ Date ___________________

Skills Worksheet

# Vocabulary and Section Summary

## Scientific Models

### VOCABULARY

**In your own words, write a definition of the following terms in the space provided.**

1. model

_______________________________________________________________

_______________________________________________________________

2. theory

_______________________________________________________________

_______________________________________________________________

3. law

_______________________________________________________________

_______________________________________________________________

### SECTION SUMMARY

**Read the following section summary.**

- A model uses familiar things to describe unfamiliar things.
- Physical, mathematical, and conceptual models are commonly used in science.
- A scientific theory is an explanation for many hypotheses and observations.
- A scientific law summarizes experimental results and observations. It describes what happens, but not why.

Name________________________________ Class __________________ Date __________________

Skills Worksheet

# Vocabulary and Section Summary

## Science and Engineering

### VOCABULARY

**In your own words, write a definition of the following terms in the space provided.**

1. technology

___________________________________________________________________________

___________________________________________________________________________

2. engineering

___________________________________________________________________________

___________________________________________________________________________

3. engineering design process

___________________________________________________________________________

___________________________________________________________________________

4. prototype

___________________________________________________________________________

___________________________________________________________________________

5. cost benefit analysis

___________________________________________________________________________

___________________________________________________________________________

6. bioengineering

___________________________________________________________________________

___________________________________________________________________________

7. assistive bioengineering

___________________________________________________________________________

___________________________________________________________________________

8. adaptive bioengineering

___________________________________________________________________________

___________________________________________________________________________

## SECTION SUMMARY

**Read the following section summary.**

- Science, technology, engineering, and mathematics are closely related.
- Engineers develop technologies to meet social, political, and economic needs.
- The engineering design process describes the steps for developing new technologies.
- Technology has intended benefits and unintended consequences.
- Bioengineering is engineering that develops technology for living things.
- Assistive bioengineering develops technologies that assist but do not change living things.
- Adaptive bioengineering develops technologies that help living things by changing them.

Name________________________________Class __________________ Date __________________

Skills Worksheet

# Vocabulary and Section Summary

## Tools, Measurement, and Safety

### VOCABULARY

**In your own words, write a definition of the following terms in the space provided.**

1. meter

_______________________________________________________________

_______________________________________________________________

2. area

_______________________________________________________________

_______________________________________________________________

3. mass

_______________________________________________________________

_______________________________________________________________

4. volume

_______________________________________________________________

_______________________________________________________________

5. density

_______________________________________________________________

_______________________________________________________________

6. temperature

_______________________________________________________________

_______________________________________________________________

### SECTION SUMMARY

**Read the following section summary.**

- Tools are used to make observations, take measurements, and analyze data.
- The International System of Units (SI) is the standard system of measurement.
- Length, volume, mass, and temperature are quantities of measurement.
- Density is the amount of matter in a given volume.
- Safety symbols are for your protection.

Name________________________ Class ______________ Date ______________

Skills Worksheet

# Directed Reading A

## Section: Sorting It All Out

1. What is classification?

_______________________________________________

_______________________________________________

### WHY CLASSIFY?

_____ 2. Putting plants and animals into orderly groups based on similar characteristics is called
a. arrangement.
b. classification.
c. identification.
d. biology.

_____ 3. Classifying living things helps human beings
a. improve the world.
b. make sense of the world.
c. destroy the world.
d. make sense of the useful plants only.

### HOW DO SCIENTISTS CLASSIFY ORGANISMS?

_____ 4. Taxonomy is the science of
a. naming plants and animals.
b. describing, classifying, and naming organisms.
c. naming and describing living things.
d. describing organisms.

_____ 5. Today, a system of classification similar to the system developed by Carolus Linnaeus
a. includes only plants.
b. is no longer used.
c. is still used.
d. does not include plants.

_____ 6. The more closely related living things are to each other, the more
a. characteristics they share.
b. food they share.
c. space they share.
d. water they will share.

_____ 7. Organisms are thought to be closely related when they have
a. almost no characteristics in common.
b. no characteristics in common.
c. few characteristics in common.
d. many characteristics in common.

_____ 8. Bears, lions, and house cats give birth to live young, and lions and house cats have retractable claws. Which of the three types of animals are most closely related?
a. lions and house cats
b. lions and bears
c. house cats and bears
d. None of the animals are related.

9. Before the 1600s, scientists divided organisms into what two groups?

______________________________________________________________

______________________________________________________________

10. What Swedish scientist created the first organized, modern taxonomy?

______________________________________________________________

11. How many levels of classification do scientists use today?

______________________________________________________________

12. Why are the platypus, brown bear, lion, and house cat thought to be related to each other?

______________________________________________________________

______________________________________________________________

______________________________________________________________

13. What characteristics do the bear, lion, and house cat have that the platypus does not have?

______________________________________________________________

______________________________________________________________

**LEVELS OF CLASSIFICATION**

_____ 14. All organisms are classified into
a. one of three domains.
b. one of eight phyla.
c. plants or animals.
d. living or nonliving things.

_____ 15. Each domain of organisms is divided into several
a. genera.
b. classes.
c. orders.
d. kingdoms.

_____ 16. The smallest, most specific classification level is
a. phylum.
b. species.
c. class.
d. order.

17. The plural form of the word phylum is ______________________.

18. What is a group of organisms that are closely related and can mate to produce fertile offspring called?

_______________________________________________________________

19. In order from largest to smallest, what are the eight levels of classification?

_______________________________________________________________

_______________________________________________________________

_______________________________________________________________

_______________________________________________________________

_______________________________________________________________

_______________________________________________________________

_______________________________________________________________

_______________________________________________________________

**SCIENTIFIC NAMES**

20. No matter how many common names an organism might have, it only has one

_______________________________________________________________

21. How was the naming of organisms different before Carolus Linnaeus, and how was the system difficult for scientists?

_______________________________________________________________

_______________________________________________________________

_______________________________________________________________

_______________________________________________________________

22. Who simplified the naming of living things by giving each species a two-part scientific name?

_______________________________________________________________

23. In the scientific name for the Asian elephant, *Elephas maximus*, the word *Elephas* indicates the animal's ____________________.

24. All genus names begin with a(n) ____________________.

25. All specific names begin with a(n) ____________________.

26. Scientific names are usually in one of these two languages,

_______________________________________________

_______________________________________________

27. In the scientific name *Tyrannosaurus rex,* what is the species name?

_______________________________________________

28. What abbreviation do scientists sometimes use when referring to *Tyrannosaurus rex*?

_______________________________________________

29. What is the scientific name for the common house cat?

_______________________________________________

**DICHOTOMOUS KEYS**

_____ 30. Scientists use dichotomous keys to

a. name organisms.
b. count organisms.
c. identify organisms.
d. catch organisms.

31. What kind of identification aid are scientists using when they work through a series of paired, descriptive statements?

_______________________________________________

**A GROWING SYSTEM**

_____ 32. Of all the organisms on the Earth,

a. all have been discovered.
b. all have been classified.
c. not all have been discovered or classified.
d. all have been given scientific names.

_____ 33. What do scientists do when a newly discovered organism does not fit any existing category?

a. leave the organism alone
b. try to change the organism
c. destroy the organism
d. create a new category

_____ 34. What newly discovered organism, first found about 50 years ago, did not fit in any existing phyla?

a. *Xenoturbella*
b. *Felis domesticus*
c. *Elephas maximus*
d. *Tyrannosaurus rex*

Name______________________________ Class ________________ Date ________________

Skills Worksheet

# Directed Reading A

## Section: Domain and Kingdoms

1. Before the discovery of organisms like *Euglena,* how were all organisms classified?

______________________________________________________________

______________________________________________________________

### WHAT IS IT?

_____ 2. Scientists classify organisms based on their what?
a. shape c. size
b. smell d. characteristics

_____ 3. Which characteristic is *not true* for euglenoids including the genus *Euglena?*
a. single celled c. live in pond water
b. live in salt water d. make their own food

_____ 4. A green color and the ability to make food through photosynthesis might make some people think that members of the genus *Euglena* are
a. trees. c. plants.
b. algae. d. mosses.

_____ 5. Which is a characteristic that animals and members of the genus *Euglena* possess but plants do not?
a. ability to move by themselves c. ability to take in water
b. ability to make food d. ability to use energy

6. What kingdom did scientists add to create a classification for organisms that had characteristics of both plants and animals?

______________________________________________________________

7. Today, there are ______________________ domains in the classification system.

### THE DOMAIN ARCHAEA

8. Single-celled organisms that do not have a nucleus are called

______________________.

9. How are archaea distinguished from other prokaryotes?

______________________________________________________________

**Identify the correct bacteria kingdom for the organisms described below by writing Archaea or Bacteria in the space provided.**

_______________________10. Some of these live inside humans.

_______________________11. One of these causes pneumonia.

_______________________12. These live in places where most other organisms could not live.

_______________________13. Its name comes from a word that means "ancient."

_______________________14. One type turns milk into yogurt.

## THE DOMAIN BACTERIA

15. Prokaryotes that usually have a cell wall and that usually reproduce by cell division belong to the domain _______________________.

## THE DOMAIN EUKARYA

16. All organisms whose cells have a nucleus and membrane-bound organelles are called _______________________.
17. All eukaryotes belong to the domain _______________________.
18. Members of the kingdom Protista are called _______________________.
19. Protists that have animal-like characteristics are called _______________________.
20. Protists that have plantlike characteristics are called _______________________.
21. Unlike plants, fungi do not use _______________________.
22. Unlike animals, _______________________ do not eat food.
23. How do fungi absorb nutrients from their surroundings?

_________________________________________________________________

24. Give two examples of fungi.

_________________________________________________________________

_________________________________________________________________

25. What do all members of the kingdom Plantae have in common?

_______________________________________________________________________________

_______________________________________________________________________________

_______________________________________________________________________________

26. In order for plants to make their own food through photosynthesis, they must be exposed to ______________________.

27. Where are plants found?

_______________________________________________________________________________

_______________________________________________________________________________

28. Explain why the food that plants make is important not only to the plants themselves but to other organisms as well.

_______________________________________________________________________________

_______________________________________________________________________________

_______________________________________________________________________________

29. What are two ways plants are used by other organisms?

_______________________________________________________________________________

_______________________________________________________________________________

_______________________________________________________________________________

30. What characteristics do most members of kingdom Animalia share?

_______________________________________________________________________________

_______________________________________________________________________________

31. Members of kingdom Animalia have specialized sense organs that allow them to respond to their ______________________.

32. Members of kingdom Animalia are commonly called ______________________.

33. Explain why animals need plants.

_______________________________________________________________________________

_______________________________________________________________________________

34. Explain how animals depend on bacteria and fungi.

________________________________________________________________________________

________________________________________________________________________________

**STRANGE ORGANISMS**

35. The kingdom Animalia includes some very simple animals, such as

    ____________________________, that do not have sense organs and cannot move.

Name________________________________Class __________________ Date __________________

Skills Worksheet

# Vocabulary and Section Summary

## Sorting It All Out

### VOCABULARY

**In your own words, write a definition of the following terms in the space provided.**

1. classification

_______________________________________________________________

_______________________________________________________________

2. taxonomy

_______________________________________________________________

_______________________________________________________________

3. dichotomous key

_______________________________________________________________

_______________________________________________________________

### SECTION SUMMARY

**Read the following section summary.**

- In classification, organisms are grouped according to the characteristics the organisms share. Classification lets scientists answer important questions about the relationship between organisms.
- The eight levels of classification are domain, kingdom, phylum, class, order, family, genus, and species.
- An organism has one two-part scientific name.
- A dichotomous key is a tool for identifying organisms that uses a series of paired descriptive statements.

Name________________________________ Class __________________ Date __________________

Skills Worksheet

# Vocabulary and Section Summary

## Domains and Kingdoms

### VOCABULARY

**In your own words, write a definition of the following terms in the space provided.**

1. Archaea

______________________________________________________________________________

______________________________________________________________________________

2. Bacteria

______________________________________________________________________________

______________________________________________________________________________

3. Eukarya

______________________________________________________________________________

______________________________________________________________________________

4. Protista

______________________________________________________________________________

______________________________________________________________________________

5. Fungi

______________________________________________________________________________

______________________________________________________________________________

6. Plantae

______________________________________________________________________________

______________________________________________________________________________

7. Animalia

______________________________________________________________________________

______________________________________________________________________________

## SECTION SUMMARY

**Read the following section summary.**

- In the past, organisms were classified as plants or animals. As scientists discovered more species, they found that organisms did not always fit into one of these two categories, so they changed the classification system.
- Today, domains are the largest groups of related organisms. The three domains are Archaea and Bacteria, both of which consist of prokaryotes, and Eukarya, which consists of eukaryotes.
- The kingdoms of the domain Eukarya are Protista, Fungi, Plantae, and Animalia.

Name_______________________________Class _________________ Date _________________

Skills Worksheet

# Directed Reading A

## Section: Animal Reproduction

1. Why is it necessary for members of a species to reproduce?

_______________________________________________________________

_______________________________________________________________

### ASEXUAL REPRODUCTION

2. In ______________________, a single parent produces offspring.
3. An offspring produced by asexual reproduction is ______________________ to its parent.
4. What are two advantages of asexual reproduction?

_______________________________________________________________

_______________________________________________________________

_______________________________________________________________

5. When a part of a parent organism pinches off and forms a new organism, it is called ______________________.
6. Part of an organism breaks off, and new parts regenerate to form a new individual in the kind of asexual reproduction called ______________________.
7. Some organisms, such as the sea star, have the ability to ______________________ body parts.

### SEXUAL REPRODUCTION

_____ 8. In sexual reproduction, two parents each contribute part of the offspring's
a. genetic information.
b. upbringing.
c. eggs.
d. birth.

_____ 9. What is a female sex cell called?
a. zygote
b. egg
c. gene
d. sperm

_____ 10. What is a male sex cell called?
a. egg
b. DNA
c. sperm
d. zygote

_____ 11. The union of a sperm with an egg is called
a. fragmentation.
b. regeneration.
c. fertilization.
d. courting.

_____ 12. A fertilized egg is called a
a. zygote.
b. sperm.
c. cyclops.
d. gene.

_____ 13. Where is genetic information found?
a. in atoms
b. in birth
c. in genes
d. in food

_____ 14. What is the location of genes?
a. on chromosomes made of protein and DNA
b. in special molecules made of eggs and sperm
c. on chromosomes made of fats and protein
d. in special molecules made of fats and DNA

15. The combination of genes from two parents results in a zygote that grows into a(n) ______________________.

16. What does the combination of genes in sexual reproduction allow for?

________________________________________________________________

17. Why is variation of genes an advantage of sexual reproduction?

________________________________________________________________

________________________________________________________________

**INTERNAL AND EXTERNAL FERTILIZATION**

_____ 18. What occurs during external fertilization?
a. The sperm fertilizes eggs outside the female's body.
b. The male produces eggs and sperm.
c. The male puts sperm inside the female.
d. The female produces both eggs and sperm.

_____ 19. Why must external fertilization take place in a moist environment?
a. so that the eggs are protected
b. so that the zygotes won't dry out
c. because the environment is warmer
d. because the environment is cooler

_____ 20. Which of the following is an advantage of internal fertilization?
a. More offspring are produced.
b. It is easier to find a mate.
c. The males produce fewer sperm.
d. The fertilized eggs are protected.

_____ 21. Most mammals give birth to live young that develop
a. inside an egg.
b. inside the female's body.
c. inside a pouch.
d. in a dry environment.

**MAMMALS**

_____ 22. All mammals reproduce sexually and feed their young
a. milk.
b. seeds.
c. eggs.
d. blood.

**Match the correct description with the correct term. Write the letter in the space provided.**

_____ 23. The female nourishes young inside her body before giving birth to well-developed newborns.

_____ 24. The female gives birth to partially developed young that continue to develop inside the mother's pouch.

_____ 25. The female lays eggs that hatch, then nourishes her young with milk.

a. monotreme
b. placental mammal
c. marsupial

**For each type of mammal listed, write whether it is a marsupial, a placental mammal, or a monotreme.**

______________________ 26. human

______________________ 27. platypus

______________________ 28. opossum

______________________ 29. echidna

______________________ 30. kangaroo

______________________ 31. armadillo

Name_______________________________ Class __________________ Date __________________

Skills Worksheet

# Directed Reading A

## Section: Plant Reproduction

1. What are the two stages in a plant's life cycle?

_______________________________________________________________

_______________________________________________________________

2. What happens during each stage of a plant's life cycle?

_______________________________________________________________

_______________________________________________________________

_______________________________________________________________

3. What must happen in order for plants to reproduce sexually?

_______________________________________________________________

### REPRODUCTION IN NONVASCULAR PLANTS

_____ 4. Where do eggs and sperm form in nonvascular plants?
  a. in sporophytes
  b. in flowers
  c. in separate structures
  d. in one special structure

_____ 5. What happens when water covers clumps of nonvascular gametophytes?
  a. Sperm swim to the female gametophytes and fertilize the eggs.
  b. The eggs float to the sperm and are fertilized.
  c. The eggs and sperm are drowned before they can meet.
  d. The eggs and sperm change back into sporophytes.

6. Why do nonvascular plants produce a large number of spores?

_______________________________________________________________

_______________________________________________________________

### REPRODUCTION IN SEEDLESS VASCULAR PLANTS

_____ 7. How are seedless vascular plants and nonvascular plants similar?
  a. Both have very large gametophytes.
  b. In both, eggs and sperm are always produced on the same plant.
  c. In both, the gametophytes change into spores.
  d. Both need water in order to reproduce.

## REPRODUCTION IN SEED PLANTS

_____ 8. Trees and shrubs that do not have flowers or fruit are
a. angiosperms.
b. sporophytes.
c. gametophytes.
d. gymnosperms.

_____ 9. Which of the following allows seed plants to live in many places?
a. Most can reproduce sexually without water.
b. Most need water to reproduce.
c. Most do not have fruit.
d. Most have cones.

_____ 10. Most gymnosperms have reproductive structures called
a. fruit.
b. cones.
c. flowers.
d. stems.

_____ 11. The pollen of a plant contains
a. eggs.
b. cones.
c. grains.
d. sperm.

_____ 12. Wind transfers pollen from the male cone to the female cone during
a. mating
b. fertilization
c. pollination
d. flowering

_____ 13. Where do gametophytes grow in angiosperms?
a. within flowers
b. within cones
c. within petals
d. within leaves

**Match the labels to the parts of the drawing. Write the letters in the spaces provided. Some labels may be used more than once.**

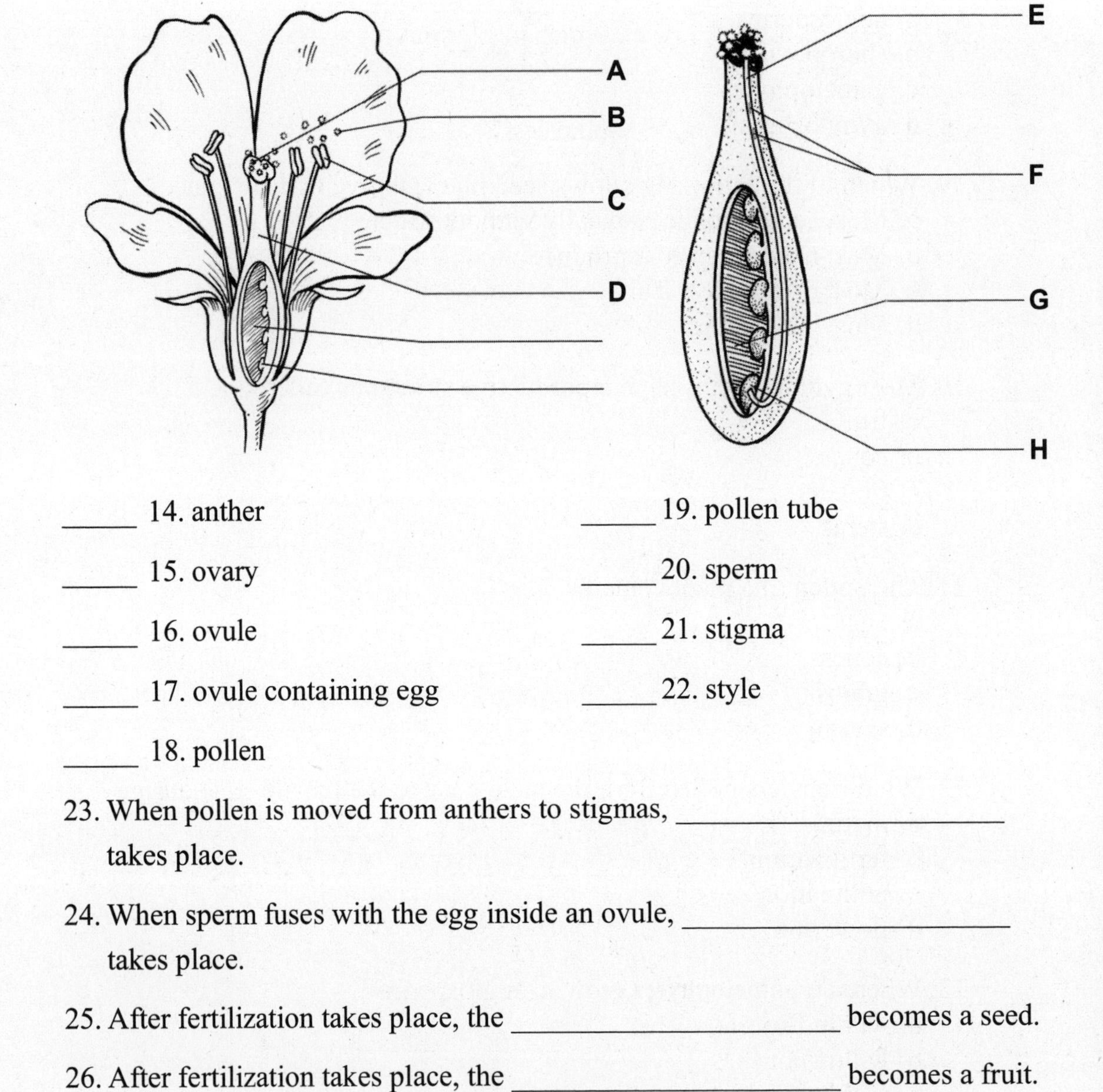


_____ 14. anther

_____ 15. ovary

_____ 16. ovule

_____ 17. ovule containing egg

_____ 18. pollen

_____ 19. pollen tube

_____ 20. sperm

_____ 21. stigma

_____ 22. style

23. When pollen is moved from anthers to stigmas, ____________________ takes place.

24. When sperm fuses with the egg inside an ovule, ____________________ takes place.

25. After fertilization takes place, the ____________________ becomes a seed.

26. After fertilization takes place, the ____________________ becomes a fruit.

27. What does a fruit do as it swells and ripens?

______________________________________________________________

28. What three things do seeds need in order to sprout?

______________________________________________________________

29. What is the advantage of asexual reproduction in angiosperms?

______________________________________________________________

______________________________________________________________

**Match the correct definition with the correct term. Write the letter in the space provided.**

_____ 30. above-ground stems from which new plants can grow

_____ 31. tiny plants that grow along the edges of a plant's leaves and fall off and grow on their own

_____ 32. underground stems from which new plants can grow

a. plantlets
b. tubers
c. runners

Name______________________________Class ________________ Date ________________

Skills Worksheet

# Directed Reading A

## Section: Animal Behavior

### KINDS OF BEHAVIOR

_____ 1. Behavior that doesn't depend on learning or experience is called
a. migration behavior.
b. territory behavior.
c. innate behavior.
d. learned behavior.

_____ 2. Innate behaviors are inherited through
a. learning.
b. genes.
c. colorful objects.
d. predators.

_____ 3. A newborn whale's ability to swim is an example of an innate behavior that
a. is not present at birth.
b. develops months or years after birth.
c. is learned.
d. is present at birth.

_____ 4. A one-year-old baby's first steps are an example of an innate behavior that
a. develops some time after birth.
b. is present at birth.
c. is learned.
d. is tied to learning language.

5. Behavior that has been learned from experience or from observing other animals is called ______________________ behavior.

### SURVIVAL BEHAVIOR

6. Name five behaviors that animals must use to survive.

_______________________________________________________________

_______________________________________________________________

_______________________________________________________________

_______________________________________________________________

_______________________________________________________________

7. Animals that eat other animals are called ______________________.

8. The animals that are eaten are called ______________________.

9. Give an example of an animal that can be both predator and prey.

_______________________________________________

_______________________________________________

_______________________________________________

10. A(n) ______________________ is an area that is occupied by one animal or by a group of animals that do not allow other members of the species to enter.

11. Claiming a territory allows animals to save energy by avoiding ______________________.

12. What are three activities animals use their territory for?

_______________________________________________

_______________________________________________

_______________________________________________

13. Name three things animals defend from other animals.

_______________________________________________

_______________________________________________

_______________________________________________

14. Describe three examples of defensive behaviors that animals may use against other animals.

_______________________________________________

_______________________________________________

_______________________________________________

15. The special behaviors that animals use to help them find mates are referred to as ______________________.

16. Many young animals depend on their ______________________ for survival.

17. Some adult birds bring ______________________ to their young.

18. Adult killer whales spend years teaching their young how to ______________________.

## SEASONAL BEHAVIOR

**Match the correct behavior with the correct animals. Write the letter in the space provided.**

_____ 19. hide from the cold by burrowing in the mud

_____ 20. store food to prepare for the winter

_____ 21. migrate to central Mexico to wait for the spring

_____ 22. fly south thousands of kilometers for the winter

_____ 23. hibernate to deal with winter food and water shortages

_____ 24. estivate to deal with summer food and water shortages

a. birds in the Northern Hemisphere
b. bears
c. frogs
d. monarch butterflies
e. desert squirrels and desert mice
f. squirrels

25. To ________________________ is to travel from one place to another.

26. What are four reasons animals may migrate?

______________________________________________________________________

______________________________________________________________________

______________________________________________________________________

______________________________________________________________________

27. Fixed objects that animals use to find their way along migration paths are called ________________________.

28. Hibernation is a period of ________________________ and decreased ________________________ that some animals experience in winter.

29. Hibernating animals survive on stored.

30. Describe what happens to an animal's body during hibernation.

______________________________________________________________________

______________________________________________________________________

______________________________________________________________________

______________________________________________________________________

31. Do bears hibernate? Explain your answer.

______________________________________________________________

______________________________________________________________

______________________________________________________________

______________________________________________________________

32. A period of reduced activity that some animals undergo in the summer is called ____________________.

33. Give two reasons why animals need to keep track of time.

______________________________________________________________

______________________________________________________________

34. The internal control of an animal's natural cycles is called a(n) ____________________.

35. What are two clues animals may use to set their internal clocks?

______________________________________________________________

______________________________________________________________

36. Some animals have biological clocks that keep track of daily cycles called ____________________.

37. Name two behaviors that are controlled by seasonal cycles.

______________________________________________________________

______________________________________________________________

38. How does reproducing during a particular season help an animal?

______________________________________________________________

______________________________________________________________

______________________________________________________________

______________________________________________________________

Name________________________________Class ________________ Date ________________

Skills Worksheet

# Directed Reading A

## Section: Adaptations and Survival

1. What is an adaptation?

______________________________________________________________

______________________________________________________________

2. Why couldn't a sea turtle survive for very long in a land environment?

______________________________________________________________

______________________________________________________________

______________________________________________________________

### ADAPTATIONS FOR OBTAINING FOOD

_____ 3. The chameleon's long, fast-moving tongue is an example of an adaptation
a. for swimming.
b. for obtaining food.
c. for fighting off predators.
d. for attracting a mate.

_____ 4. What is an example of an adaptation that helps humans obtain food?
a. strong teeth and jaws
b. the ability to drive cars
c. the shape and function of their hands and fingers
d. the ability to attract a mate

### PREDATOR-PREY ADAPTATIONS

_____ 5. Many organisms have adaptations that
a. make it hard for them to survive.
b. make it hard for them to reproduce.
c. make it easier for prey to see them.
d. serve as a defense against predators.

_____ 6. What does camouflage do for an organism?
a. It disguises the organism so it is hard to see.
b. It helps the organism to attract a mate.
c. It lets the organism breathe under water.
d. It warns predators to leave the organism alone.

7. Why are patterns with black stripes and red, orange, or yellow markings common in dangerous animals such as poisonous frogs and snakes?

_______________________________________________________________

_______________________________________________________________

8. Why are some predators camouflaged by their environment?

_______________________________________________________________

_______________________________________________________________

## ADAPTATIONS TO INTERACTIONS

_____ 9. Two species can adapt to
a. reproduce together.
b. eat with one another.
c. fight with one another.
d. interact with one another.

_____ 10. Which of the following is NOT true of adaptations based on interactions between organisms?
a. They happen over a very long period of time.
b. They happen over a very short period of time.
c. They happen between organisms living close together.
d. They may happen between plant and animal species.

11. Describe how the honeycreeper and the lobelia plant are adapted to one another.

_______________________________________________________________

_______________________________________________________________

_______________________________________________________________

_______________________________________________________________

## NATURAL SELECTION

_____ 12. What explains the change in the inherited characteristics of a population over time?
a. interaction
b. natural selection
c. survival
d. reproduction

13. Define *natural selection.*

_______________________________________________

_______________________________________________

_______________________________________________

14. What may influence which baby animals, such as sea turtles, grow to adulthood and which do not?

_______________________________________________

_______________________________________________

15. If an environment changes, which individuals are likely to survive and reproduce?

_______________________________________________

_______________________________________________

**Match the correct description with the correct step of natural selection. Write the letter in the space provided.**

_____ 16. More offspring are born than will live to become adults.

_____ 17. The individuals in a population are different from one another.

_____ 18. Many individuals will die before reproducing.

_____ 19. The individuals best adapted to their environment will live to reproduce.

a. genetic variation within a population
b. successful reproduction
c. overproduction
d. struggle to survive

20. What is the value of genetic variation within a population?

_______________________________________________

_______________________________________________

_______________________________________________

21. What are three factors that might kill off less well-adapted individuals struggling to survive in a population?

_______________________________________________

_______________________________________________

## CHANGES IN GENETIC VARIATION

_____ 22. Over time, a population will be made up of more individuals with characteristics that will
a. help them survive.
b. not help them survive.
c. keep them from reproducing.
d. make them attractive to predators.

_____ 23. In order for natural selection to continue, individuals within a population must be
a. genetically identical.
b. genetically different.
c. genetically altered.
d. genetically enhanced.

24. If a rapid decrease occurs in a population, what may happen to the characteristics of that population? Explain your answer.

_______________________________________________________________

_______________________________________________________________

25. If a population is able to increase after a genetic bottleneck, what will the individuals be like?

_______________________________________________________________

_______________________________________________________________

26. How will genetic similarity affect a population?

_______________________________________________________________

_______________________________________________________________

27. What are three reasons for the loss of genetic variation in the Florida panther population?

_______________________________________________________________

_______________________________________________________________

_______________________________________________________________

**INSECTICIDE RESISTANCE**

_____ 28. What is insecticide resistance a result of?
a. natural selection
b. a genetic bottleneck
c. competition
d. inbreeding

_____ 29. Some insects are resistant to insecticides because
a. they know how to avoid insecticides.
b. they have genes that make them resistant.
c. the insecticides help them to reproduce.
d. most insecticides don't work very well.

30. Why does an insect population include more insecticide-resistant individuals each time the same insecticide is used?

______________________________________________________________

______________________________________________________________

Name________________________________ Class __________________ Date __________________

Skills Worksheet

# Vocabulary and Section Summary

## Animal Reproduction

### VOCABULARY

**In your own words, write a definition of the following terms in the space provided.**

1. asexual reproduction

_______________________________________________________________

_______________________________________________________________

2. sexual reproduction

_______________________________________________________________

_______________________________________________________________

3. external fertilization

_______________________________________________________________

_______________________________________________________________

4. internal fertilization

_______________________________________________________________

_______________________________________________________________

### SECTION SUMMARY

**Read the following section summary.**

- In asexual reproduction, a single parent's offspring are genetically identical to the parent.
- In sexual reproduction, offspring receive a combination of genes from two parents. Combining genes allows for variation. Variation allows a population to adapt to changes.
- Fertilization can be external or internal.
- All mammals reproduce sexually.

Name______________________________ Class ________________ Date ________________

Skills Worksheet

# Vocabulary and Section Summary

## Plant Reproduction

### VOCABULARY

**In your own words, write a definition of the following term in the space provided.**

1. pollination

_______________________________________________

_______________________________________________

### SECTION SUMMARY

**Read the following section summary.**

- Most gametophytes of nonvascular plants and seedless vascular plants need water to reproduce.
- Seed plants do not need water to reproduce. During fertilization, wind and animals can transfer pollen from cone to cone or from flower to flower.
- Some plants use plantlets, tubers, or runners to reproduce asexually.

Name________________________ Class ______________ Date ______________

Skills Worksheet

# Vocabulary and Section Summary

## Animal Behavior

### VOCABULARY

**In your own words, write a definition of the following terms in the space provided.**

1. innate behavior

___________________________________________

___________________________________________

2. learned behavior

___________________________________________

___________________________________________

3. territory

___________________________________________

___________________________________________

4. hibernation

___________________________________________

___________________________________________

5. estivation

___________________________________________

___________________________________________

### SECTION SUMMARY

**Read the following section summary.**

- Behavior may be classified as innate or learned. The potential for innate behavior is inherited. Learned behavior depends on experience.
- Behaviors that help animals survive include finding food, marking a territory, defensive action, courtship, and parenting.
- Seasonal behaviors, such as hibernation and estivation can help animals adapt to the environment.

Name________________________________ Class __________________ Date __________________

Skills Worksheet

# Vocabulary and Section Summary

## Adaptations and Survival

### VOCABULARY

**In your own words, write a definition of the following terms in the space provided.**

1. adaptation

_______________________________________________________________

_______________________________________________________________

2. natural selection

_______________________________________________________________

_______________________________________________________________

### SECTION SUMMARY

**Read the following section summary.**

- A strong beak, bright colors, and camouflage are adaptations that can help an organism survive.
- The four parts of natural selection are overproduction, genetic variation, competition for resources, and successful reproduction.
- Variation is due to the exchange of genetic information as it is passed from parent to offspring.
- Genetic variation allows a population to adapt to changes in the environment over time.
- When a population of insects is resistant to an insecticide, survival rates can increase.

Name______________________________ Class ________________ Date ________________

Skills Worksheet

# Directed Reading A

## Section: Change Over Time

1. One way to tell kinds of animals apart is by their ____________________.

### DIFFERENCES BETWEEN ORGANISMS

_____ 2. How does adaptation help an organism?
a. It helps the organism change colors.
b. It improves its ability to survive and reproduce.
c. It improves its ability to change species.
d. It helps the organism become a fossil.

_____ 3. If one animal or plant has the same characteristics as another, they may both be part of the same
a. organism.
b. planet.
c. species.
d. fossil record.

4. Two organisms that can mate to produce offspring that can reproduce belong to the same ____________________.

5. When members of the same species live in the same place, they form a(n) ____________________.

6. Since life began on Earth, many ____________________ have vanished and many new ones have appeared.

7. Scientists have observed that species ____________________ over time.

8. The inherited ____________________ in populations also change over time.

9. What can result as populations of organisms change?

______________________________________________________________

### EVIDENCE OF CHANGES OVER TIME

_____ 10. Where do scientists look for evidence of change over time?
a. in the layers of Earth's crust
b. in igneous rock
c. in water
d. in old books

_____ 11. What is a fossil?
  a. a layer of sediment
  b. evidence of a living organism
  c. remains of a still-living organism
  d. remains of a once-living organism

12. Describe how a fossil is usually formed.

_______________________________________________

_______________________________________________

_______________________________________________

_______________________________________________

_______________________________________________

13. What is the timeline of life formed by studying fossils called?

_______________________________________________

14. How are fossils organized in the fossil record?

_______________________________________________

_______________________________________________

15 Fossils in newer layers of the Earth tend to resemble present-day

________________________.

16. In older layers of the Earth, are fossils more or less likely to resemble today's animals or plants?

_______________________________________________

17. Some fossils may be of earlier life-forms that do not

________________________ anymore.

## EVIDENCE OF ANCESTRY

_____ 18. The fossil record provides evidence about
  a. the age of rocks.
  b. the order in which species have existed.
  c. the number of layers the Earth has.
  d. the composition of minerals.

_____ 19. All living things inherit characteristics in similar ways from their
  a. ancestors.
  b. environment.
  c. fossils.
  d. descendants.

20. As scientists study the fossil record, they may draw models to illustrate their

    ______________________ about the relationships between extinct and living

    organisms.

21. How does the scientists' model indicate a group of organisms that descended from a species?

    ______________________________________________________________________

22. List two groups of animals that may share a common ancestor with whales.

    ______________________________________________________________________

    ______________________________________________________________________

23. Scientists think that all mammal species alive today arose from common

    ______________________.

24. Scientists have combined information on hundreds of thousands of organisms

    to sketch out a ______________________ that includes all known organisms.

25. What does the lack of a fossil record for some of the Earth's history mean to scientists?

    ______________________________________________________________________

**EXAMINING ORGANISMS**

26. In addition to fossils, how can scientists learn about an organism's ancestors?

    ______________________________________________________________________

27. List three things about whales that tell scientists that they are not fish.

    ______________________________________________________________________

    ______________________________________________________________________

    ______________________________________________________________________

28. What does a whale body have that hints it had an ancestor that lived on land?

    ______________________________________________________________________

    ______________________________________________________________________

## COMPARING ORGANISMS

_____ 29. What do organisms inherit from ancestors?
a. mammal characteristics
b. traits determined by DNA
c. traits caused by environment
d. new traits

_____ 30. What makes the human hand similar to a dolphin's flipper or a bat's wing?
a. the ability to flap
b. the structure and composition of the skin
c. the order of their changes over time
d. the structure and order of bones

_____ 31. What does the similarity between humans, dolphins, cats, and bats indicate?
a. that they all appeared recently
b. that their ancestors lived in the same place
c. that they share a common ancestor
d. that they are becoming more alike over time

_____ 32. If organisms with similar traits arise from a common ancestor, what will they share?
a. similar DNA
b. similar arms and legs
c. the ability to mate with each other
d. similar fossils

Name_______________ Class _______________ Date _______________

Skills Worksheet

# Directed Reading A

## Section: How Do Population Changes Happen?

1. List three things that scientists learned about Earth beginning in the 1800s.

_______________

_______________

_______________

### CHARLES DARWIN

_____ 2. What did Darwin do in order to study plants and animals?
   a. He took a trip around the world.
   b. He studied theology.
   c. He formed theories.
   d. He became a doctor.

_____ 3. What did Darwin do during his travels?
   a. He wrote a book about his theory.
   b. He collected thousands of plant and animal samples.
   c. He took photos of plants and animals.
   d. He visited all the continents.

4. Darwin noticed that the plants and animals on the _______________ were similar to, but not the same as, those in Ecuador.

5. What was one way that finches on different islands differed from each other?

_______________

6. What was the beak of each finch adapted to?

_______________

### DARWIN'S THINKING

_____ 7. What puzzled Darwin about the Galápagos finches?
   a. They were so different.
   b. They should not have been there.
   c. They were too similar.
   d. They were similar, but had many differences.

_____ 8. A form of a hereditary characteristic is a(n)
   a. species.
   b. breeding.
   c. trait.
   d. adaptation.

9. What hypothesis did Darwin develop about the Galápagos finches?

______________________________________________________________

______________________________________________________________

______________________________________________________________

______________________________________________________________

**Match the correct definition with the correct term. Write the letter in the space provided.**

_____ 10. the idea that human populations can grow faster than the food supply

_____ 11. the idea that Earth had formed naturally over a long period of time

_____ 12. the practice of breeding plants and animals to have desired traits

a. Lyell's theory
b. selective breeding
c. Malthus's principle

13. Why do farmers and breeders use selective breeding?

______________________________________________________________

______________________________________________________________

14. After reading Malthus's theory, Darwin realized that any species can produce many ______________________________ .

15. The number of a species' offspring is limited by starvation, disease, predation, or ______________________.

16. Darwin reasoned that species could inherit traits that would help them ______________________ in their environment.

17. Darwin had begun to think that species could ______________________ over time.

18. What idea became clear to Darwin after he read Principles of Geology, by Charles Lyell?

______________________________________________________________

______________________________________________________________

**DARWIN'S THEORY OF NATURAL SELECTION**

19. What was the name of Darwin's famous book?

_______________________________________________

20. What was Darwin's theory that organisms that are better adapted to their environment are more successful than organisms that are less well adapted called?

_______________________________________________

**Match the correct description with the correct term. Write the letter in the space provided.**

_____ 21. many more offspring are produced than will survive

_____ 22. no two offspring are alike

_____ 23. many offspring will be killed before reproducing

_____ 24. the best adapted organisms will have many offspring

a. inherited variation
b. struggle to survive
c. overproduction
d. successful reproduction

25. List two things that Darwin did not know in relation to his theory.

_______________________________________________

_______________________________________________

_______________________________________________

26. Today, scientists know that variation happens as a result of the exchange of ____________________ as it is passed from parent to offspring.

27. When organisms carry genes that make them more likely to survive to reproduce, the process called ____________________ occurs.

Name______________________________ Class ________________ Date ________________

Skills Worksheet

# Directed Reading A

## Section: Natural Selection in Action

_____ 1. Bacteria passing resistance to a medicine on to offspring is an example of

a. natural selection.
b. chemical action.
c. genetic change.
d. overproduction.

### CHANGES IN POPULATIONS

_____ 2. What does natural selection explain about a population?

a. how long it has been since it has changed
b. how it changes in response to its environment
c. how it resists change
d. how likely its members are to leave fossils

_____ 3. Which individuals in a population are most likely to survive and reproduce?

a. the largest ones
b. the ones with the most DNA
c. the well-adapted ones
d. the oldest ones

_____ 4. The growing rate of tuskless elephants in Uganda is an example of

a. selective breeding.
b. luck.
c. adaptation.
d. speciation.

5. Why are tuskless elephants becoming more likely to reproduce than elephants with tusks?

______________________________________________________________

______________________________________________________________

6. The ability of some insect species to resist chemicals is called insecticide ______________________.

7. The period of time between the birth of one generation and the birth of the next is known as the ______________________.

8. Insect species can become resistant quickly because they have a short ______________________.

9. Survival is only a part of natural selection. The other part takes place when organisms ______________________.

10. When competition for mates is intense, many organisms select mates for

    ________________________.

## FORMING A NEW SPECIES

**Match the correct description with the correct term. Write the letter in the space provided.**

_____ 11. the formation of new species

_____ 12. changes in response to the environment

_____ 13. the loss of ability of separated groups to interbreed

_____ 14. the moving apart of populations

a. adaptation
b. division
c. separation
d. speciation

15. Describe the process of forming a new species after separation.

    ______________________________________________________________________

    ______________________________________________________________________

    ______________________________________________________________________

    ______________________________________________________________________

16. When a portion of a population becomes isolated, ________________________ often begins.
17. Through adaptation, members of separated groups may eventually have different ________________________.
18. If environmental conditions differ, ________________________ will also differ.
19. When members of related groups can no longer interbreed, they have become members of different ________________________.

Name______________________________Class ________________ Date ________________

Skills Worksheet

# Vocabulary and Section Summary

## Change Over Time

### VOCABULARY

**In your own words, write a definition of the following terms in the space provided.**

1. adaptation

________________________________________________________________

________________________________________________________________

2. species

________________________________________________________________

________________________________________________________________

3. fossil

________________________________________________________________

________________________________________________________________

4. fossil record

________________________________________________________________

________________________________________________________________

### SECTION SUMMARY

**Read the following section summary.**

- The fossil record provides evidence that changes in the kinds of organisms in the environment have been occurring over time.
- Evidence that organisms change over time can be found by comparing living organisms to each other. Such comparisons provide evidence of common ancestry.
- Scientists think that modern whales arose from an ancient, land-dwelling mammal ancestor. Fossil organisms that support this hypothesis have been found.
- Comparing DNA and inherited traits provides evidence of common ancestry among living organisms. The traits and DNA of species that have a common ancestor are more similar to each other than they are to those of distantly related species.

Skills Worksheet

# Vocabulary and Section Summary

## How Do Population Changes Happen?

### VOCABULARY

**In your own words, write a definition of the following terms in the space provided.**

1. trait

______________________________________________________________________

______________________________________________________________________

2. selective breeding

______________________________________________________________________

______________________________________________________________________

3. natural selection

______________________________________________________________________

______________________________________________________________________

### SECTION SUMMARY

**Read the following section summary.**

- Finch species of the Galápagos Islands developed various adaptations in response to their environment.
- Natural selection is the process by which organisms that are better adapted to their environment are more likely to survive.
- The four parts of Darwin's theory of natural selection include overproduction, inherited variation, competition for resources, and successful reproduction.
- Variation in each species is due to the exchange of genetic information as it is passed from parent to offspring.

Name_______________________________Class ________________ Date ________________

Skills Worksheet

# Vocabulary and Section Summary

## Natural Selection in Action

### VOCABULARY

**In your own words, write a definition of the following terms in the space provided.**

1. generation time

_______________________________________________

_______________________________________________

2. speciation

_______________________________________________

_______________________________________________

### SECTION SUMMARY

**Read the following section summary.**

- Natural selection can result in an adaptation that helps an organism survive. Two such examples are the tuskless trait in elephants and insecticide resistance in insects.
- Natural selection explains how one species changes into another. Speciation occurs as populations undergo separation, adaptation, and division.

Name_______________ Class _______________ Date _______________

Skills Worksheet

# Directed Reading A

## Section: Geologic History

### THE PRINCIPLE OF UNIFORMITARIANISM

1. In his book *Theory of the Earth,* written in 1788, which scientist outlined the principle now called uniformitarianism?

_______________________________________________

2. The idea that the same geologic processes that are shaping Earth today have been at work throughout Earth's history is called _______________.

3. Describe the observations made by James Hutton that led to his development of the idea of uniformitarianism.

_______________________________________________

_______________________________________________

_______________________________________________

_______________________________________________

_______________________________________________

_______________________________________________

4. When Hutton developed his theory, how old did most people believe Earth was?

_______________________________________________

5. What observations did Hutton make at Siccar Point that supported his theory?

_______________________________________________

_______________________________________________

_______________________________________________

_______________________________________________

6. What is the principle of catastrophism?

_______________________________________________

_______________________________________________

_______________________________________________

7. What scientist caused people to seriously consider uniformitarianism as geology's guiding principle?

_______________________________________________________________

8. What book did Charles Lyell publish between 1830 and 1833, in which he reintroduced uniformitarianism?

_______________________________________________________________

**MODERN GEOLOGY—A HAPPY MEDIUM**

_____ 9. During the late 20th century, which scientist challenged Charles Lyell's principle of uniformitarianism?
a. James Hutton
b. Charles Darwin
c. Albert Einstein
d. Stephen J. Gould

_____ 10. What do scientists today believe accounts for geologic change throughout Earth's history?
a. gradual and uniform change
b. catastrophes like asteroids and comets striking Earth
c. mining and the burning of fossil fuels
d. a combination of uniformitarianism and catastrophism

_____ 11. What do scientists believe caused dinosaurs to become extinct about 65 million years ago?
a. earthquake activity
b. predators
c. an asteroid strike
d. continental drift

_____ 12. What do scientists believe happened after an asteroid struck Earth 65 million years ago?
a. A global debris cloud blocked the sun's rays for decades.
b. Debris from the impact struck and killed the dinosaurs.
c. Debris from the impact blocked the sun where the dinosaurs lived.
d. The impact blew a hole through Earth's ozone layer.

_____ 13. One of two methods used to determine the age of objects in sedimentary rocks is
a. geologic dating.
b. sedimentary layer dating.
c. estimated dating.
d. relative dating.

_____ 14. Layers of sedimentary rock have different thicknesses based on
a. the type of sediment deposited.
b. the rate at which sediment is deposited.
c. the place where sediment was deposited.
d. the number of sedimentary rock layers.

_____ 15. The top layers of sedimentary rock are usually
a. older than the bottom layers.
b. younger than the bottom layers.
c. lighter than the bottom layers.
d. heavier than the bottom layers.

16. How do scientists use a fossil's position in sedimentary rock to determine whether the fossil is older or younger than other fossils?

_______________________________________________

_______________________________________________

17. Any method of determining whether an event or object is older or younger than other events or objects is called ______________________.

18. What data have geologists combined to make the job of relative dating easier?

_______________________________________________

_______________________________________________

19. Why can rock sequences from around the world be used to create the geologic column?

_______________________________________________

20. What is the *geologic column*?

_______________________________________________

_______________________________________________

_______________________________________________

21. How do scientists use the geologic column?

_______________________________________________

**ABSOLUTE DATING**

22. Any method of measuring the age of an event or object in years is

    called ____________________________.

23. What do scientists examine in absolute dating?

    ________________________________________________________________

24. What are atoms?

    ________________________________________________________________

    ________________________________________________________________

25. What do unstable atoms release when they decay?

    ________________________________________________________________

26. What happens to an atom when it decays?

    ________________________________________________________________

    ________________________________________________________________

27. What is an atom's *half-life*?

    ________________________________________________________________

    ________________________________________________________________

28. How do scientists use atoms to determine the approximate age of a sample of rock?

    ________________________________________________________________

    ________________________________________________________________

29. How long are the half-lives of unstable atoms?

    ________________________________________________________________

    ________________________________________________________________

30. Why do scientists use uranium-238 to date rocks or fossils that are many millions of years old?

    ________________________________________________________________

    ________________________________________________________________

31. Why do scientists use carbon-14 to date fossils and other objects that are less than 50,000 years old?

    ________________________________________________________________

    ________________________________________________________________

## PALEONTOLOGY—THE STUDY OF PAST LIFE

Match the correct description with the correct term. Write the letter in the space provided.

_____ 32. the scientific study of fossils

_____ 33. scientist who studies past life

_____ 34. the remains of an organism preserved by geologic processes

_____ 35. scientist who studies the fossils of plants

a. paleontologist
b. paleobotanist
c. paleontology
d. fossil

Name________________________________ Class ___________________ Date ___________________

Skills Worksheet]

# Directed Reading A

## Section: Looking at Fossils

### FOSSILIZED ORGANISMS

_____ 1. The remains or physical evidence of an organism preserved by geologic processes is called a
a. rock.
b. mummy.
c. fossil.
d. dinosaur

_____ 2. When organisms die and are quickly buried by sediment,
a. their rate of decay speeds up.
b. their rate of decay slows.
c. the organisms do not decay.
d. the organisms decay immediately.

_____ 3. When sediments become rock,
a. hard parts of animals are preserved much more often than soft tissues are.
b. soft parts of animals are preserved much more often than hard tissues are.
c. all parts of an organism are destroyed.
d. all parts of an organism are preserved.

4. Describe how insects can be preserved in amber.

_______________________________________________________________

_______________________________________________________________

_______________________________________________________________

5. What is *amber*?

_______________________________________________________________

6. The process in which minerals replace an organism's tissues is called ________________________.

7. A form of petrifaction in which the pore space in an organism's hard tissue is filled up with mineral is called ________________________.

8. A form of petrifaction in which minerals completely replace the tissue of an organism is called ________________________.

9. Describe how organisms have been trapped and preserved in the La Brea asphalt deposits.

_______________________________________________

_______________________________________________

_______________________________________________

10. How have scientists used the fossils found in the La Brea asphalt deposits?

_______________________________________________

_______________________________________________

11. What is one type of fossil discovered frozen in the Siberian tundra?

_______________________________________________

_______________________________________________

12. Why have many fossils been found in places frozen since the last ice age?

_______________________________________________

_______________________________________________

**OTHER TYPES OF FOSSILS**

_____ 13. Any naturally preserved evidence of animal activity is called a(n)

a. trace fossil.
b. track.
c. cast.
d. mold.

_____ 14. What are two things fossilized animal tracks reveal about the animal that made them?

a. their color and shape
b. their gender and age
c. their size and speed
d. their color and gender

15. How are trace fossils of animal burrows formed?

_______________________________________________

_______________________________________________

16. A trace fossil made of preserved animal dung is called

a(n) ______________________.

17. A cavity in rock where a plant or animal was buried is called

a(n) ______________________.

18. An object created when sediment fills a mold and becomes rock is called

    a(n) ______________________.

19. What can casts show about the features of an organism?

    ______________________________________________________________

    ______________________________________________________________

**USING FOSSILS TO INTERPRET THE PAST**

20. What are four reasons that the fossil record offers only a rough sketch of the history of life on Earth?

    ______________________________________________________________

    ______________________________________________________________

    ______________________________________________________________

    ______________________________________________________________

21. What does the presence of marine fossils mean about the rocks of mountaintops in Canada?

    ______________________________________________________________

    ______________________________________________________________

22. What are two kinds of environmental change marine fossils help scientists reconstruct?

    ______________________________________________________________

    ______________________________________________________________

23. What is one thing scientists can tell from fossils of plants and land animals?

    ______________________________________________________________

    ______________________________________________________________

24. What is one way scientists study the relationships between fossils to interpret how life has changed over time?

    ______________________________________________________________

    ______________________________________________________________

25. What is one reason the fossil record is incomplete?

    ______________________________________________________________

    ______________________________________________________________

26. What are two things scientists look for to help fill in the blanks in the fossil record?

______________________________________________________________

______________________________________________________________

## USING FOSSILS TO DATE ROCKS

27. How do scientists determine the time span in which the organisms that formed particular types of fossils in certain layers of rock lived?

______________________________________________________________

______________________________________________________________

28. What does it mean if the fossils of an organism show up in a limited range of rock layers?

______________________________________________________________

______________________________________________________________

29. What are *index fossils*?

______________________________________________________________

______________________________________________________________

______________________________________________________________

30. Where must fossils be found to be index fossils?

______________________________________________________________

**Match the correct description with the correct term. Write the letter in the space provided.**

_____ 31. fossils used to establish the age of a rock layer because it is distinct, abundant, widespread, and existed for a short span of geologic time

_____ 32. genus of ammonites that lived between 230 million and 208 million years ago

_____ 33. genus of trilobites that lived approximately 400 million years ago

a. Phacops
b. index fossil
c. Tropites

34. When scientists find fossils of trilobites in rock layers, how old do they assume the rock layers are?

______________________________________________________________

Name________________________ Class ______________ Date ______________

Skills Worksheet

# Directed Reading A

## Section: Time Marches On

### GEOLOGIC TIME

_____ 1. In terms of Earth's history, how much time is 150 million years?
a. 0.3% of the time our planet has existed
b. 3% of the time our planet has existed
c. 30% of the time our planet has existed
d. 25% of the time our planet has existed

_____ 2. How much time is 4% of the time represented by Earth's oldest known rocks?
a. 1 million years
b. 1.5 million years
c. 15 million years
d. 150 million years

3. Why is Grand Canyon National Park a good place to see Earth's history recorded in rock layers?

_______________________________________________

_______________________________________________

_______________________________________________

_______________________________________________

4. How thick are the sedimentary rocks that belong to the Green River formation in Wyoming, Utah, and Colorado?

_______________________________________________

5. Why are fossils of plants and animals common in the Green River formation?

_______________________________________________

_______________________________________________

### THE GEOLOGIC TIME SCALE

_____ 6. A scale that divides Earth's 4.6 billion–year history into distinct intervals of time is called the
a. periodic time scale.
b. geologic time scale.
c. geologic time frame.
d. geologic column.

_____ 7. The eon from 542 million years ago to the present; the eon in which we live is the
a. Phanerozoic eon.
b. Hadean eon.
c. Archean eon.
d. Proterozoic eon.

_____ 8. The time from 2.5 billion years ago to 542 million years ago when the first organisms with well-developed cells appeared was the
a. Phanerozoic eon.
b. Hadean eon.
c. Archean eon.
d. Proterozoic eon.

_____ 9. The eon between 3.8 billion years ago and 2.5 billion years ago when the earliest known rocks on Earth formed was the
a. Phanerozoic eon.
b. Hadean eon.
c. Archean eon.
d. Proterozoic eon.

_____ 10. The time between 4.6 billion years ago and 3.8 billion years ago; only meteorites and moon rocks have been found from the
a. Phanerozoic eon.
b. Hadean eon.
c. Archean eon.
d. Proterozoic eon.

_____ 11. The largest division of geologic time is the
a. period.
b. era.
c. epoch.
d. eon.

_____ 12. The geologic time scale is divided into four
a. periods.
b. eras.
c. epochs.
d. eons.

_____ 13. The second-largest division of geologic time is the
a. period.
b. era.
c. epoch.
d. eon.

_____ 14. The Phanerozoic eon is divided into three
a. periods.
b. eras.
c. epochs.
d. centuries.

_____ 15. The third-largest division of geologic time is the
a. period.
b. era.
c. epoch.
d. eon.

_____ 16. Periods are divided into
- a. centuries.
- b. eras.
- c. epochs.
- d. eons.

_____ 17. The boundaries between geologic time intervals represent
- a. shorter intervals in which visible changes took place on Earth.
- b. longer intervals in which visible changes took place on Earth.
- c. short intervals in which Earth's changes cannot be seen.
- d. long intervals in which Earth's changes cannot be seen.

_____ 18. What could cause the number of species on Earth to increase dramatically?
- a. a relatively sudden increase or decrease in competition between species
- b. a gradual increase or decrease in competition between species
- c. a slow decrease in competition between species
- d. a mass extinction event

_____ 19. What could cause the number of species on Earth to dramatically decrease over a relatively short period of time?
- a. a relatively sudden increase or decrease in competition between species
- b. a gradual increase or decrease in competition between species
- c. a slow decrease in competition between species
- d. a mass extinction event

20. The death of every member of a species is called ____________________.

21. What are two gradual events that could cause mass extinctions?

_______________________________________________________________

_______________________________________________________________

22. The first era well represented by fossils is the ____________________.

23. Describe the kinds of life that appeared at the beginning, middle, and end of the Paleozoic era.

_______________________________________________________________

_______________________________________________________________

_______________________________________________________________

_______________________________________________________________

24. What brought the Paleozoic era to an end?

_______________________________________________________________

_______________________________________________________________

_______________________________________________________________

25. What do scientists believe may have caused the mass extinction of marine species at the end of the Paleozoic era?

_______________________________________________________________

_______________________________________________________________

26. Why is the Mesozoic era known as the *Age of Reptiles*?

_______________________________________________________________

_______________________________________________________________

27. What happened to 15% to 20% of the species on Earth at the end of the Mesozoic era?

_______________________________________________________________

_______________________________________________________________

28. What may have caused the extinction at the end of the Mesozoic era?

_______________________________________________________________

29. The era known as the *Age of Mammals* is the ________________________.

30. What are two unique traits that may have helped mammals survive the environmental changes that may have caused extinction of the dinosaurs?

_______________________________________________________________

_______________________________________________________________

Name________________________________Class __________________ Date __________________

Skills Worksheet

# Vocabulary and Section Summary

## Geologic History

### VOCABULARY

**In your own words, write a definition of the following terms in the space provided.**

1. uniformitarianism

_______________________________________________________________________________

_______________________________________________________________________________

2. catastrophism

_______________________________________________________________________________

_______________________________________________________________________________

3. relative age

_______________________________________________________________________________

_______________________________________________________________________________

4. relative dating

_______________________________________________________________________________

_______________________________________________________________________________

5. geologic column

_______________________________________________________________________________

_______________________________________________________________________________

6. absolute dating

_______________________________________________________________________________

_______________________________________________________________________________

7. paleontology

_______________________________________________________________________________

_______________________________________________________________________________

## SECTION SUMMARY

**Read the following section summary.**

- Uniformitarianism assumes that geologic change is gradual. Catastrophism assumes that geologic change is sudden.
- Geology used to be based on catastrophism. Modern geology is based on the idea that gradual geologic change is interrupted by catastrophes.
- Relative dating is any method of determining whether an object or event is older or younger than other objects or events.
- Absolute dating is any method of measuring the age of an object or event in years.
- Paleontology is a science that uses fossils to study past life.

Name_______________________________ Class _________________ Date _________________

Skills Worksheet

# Vocabulary and Section Summary

## Looking at Fossils

### VOCABULARY

**In your own words, write a definition of the following terms in the space provided.**

1. fossil

_______________________________________________

_______________________________________________

2. trace fossil

_______________________________________________

_______________________________________________

3. mold

_______________________________________________

_______________________________________________

4. cast

_______________________________________________

_______________________________________________

5. index fossil

_______________________________________________

_______________________________________________

### SECTION SUMMARY

**Read the following section summary.**

- Fossils are the remains or physical evidence of an organism preserved by geologic processes.
- Fossils can be preserved in rock, amber, asphalt, and ice and by petrifaction.
- Trace fossils are any naturally preserved evidence of animal activity. Tracks, burrows, and coprolites are examples of trace fossils.
- Scientists study fossils to determine how environments and organisms have changed over time.
- An index fossil is a fossil of an organism that lived for a relatively short, well-defined time span. Index fossils can be used to establish the age of rock layers.

Name______________________________Class ________________ Date ________________

Skills Worksheet

# Vocabulary and Section Summary

## Time Marches On

### VOCABULARY

**In your own words, write a definition of the following terms in the space provided.**

1. geologic time scale

_______________________________________________

_______________________________________________

2. eon

_______________________________________________

_______________________________________________

3. era

_______________________________________________

_______________________________________________

4. period

_______________________________________________

_______________________________________________

5. epoch

_______________________________________________

_______________________________________________

6. extinction

_______________________________________________

_______________________________________________

## SECTION SUMMARY

**Read the following section summary.**

- Layers of sedimentary rock form around the remains of plants and animals, creating a record of many of the organisms that have lived on Earth.
- The geologic time scale divides the 4.6 billion–year history of Earth into distinct intervals of time. The divisions of geologic time are eons, eras, periods, and epochs.
- The boundaries between geologic time intervals represent visible changes that have taken place on Earth.
- At certain times in Earth's history, the number of life-forms has increased or decreased dramatically.

Name________________________________Class __________________ Date __________________

Skills Worksheet

# Directed Reading A

## Section: Environmental Problems

1. What happened in the late 1700s that caused more harmful substances to enter the air, water, and soil?

_______________________________________________________________

_______________________________________________________________

_______________________________________________________________

### POLLUTION

______ 2. An unwanted change in the environment caused by substances or forms of energy is
a. overpopulation.
b. biodiversity.
c. pollution.
d. landfill.

______ 3. Something that causes pollution is called a
a. resource.
b. pollutant.
c. change.
d. substance.

______ 4. Pollutants can be natural, or they can be
a. unusual.
b. renewable.
c. harmful.
d. human-made.

______ 5. What does the average American throw away more of than the average person in any other nation?
a. plants
b. trash
c. animals
d. oil

______ 6. Wastes that can catch fire, eat through metal, explode, or make people sick are called
a. harmful wastes.
b. dangerous wastes.
c. hazardous wastes.
d. critical wastes.

______ 7. Which of these are hazardous wastes?
a. leftovers
b. lawn clippings
c. medical wastes
d. newspapers

8. What are some ways that chemicals are useful to people?

_______________________________________________________________

_______________________________________________________________

_______________________________________________________________

9. Describe two groups of harmful chemicals and their effects.

_______________________________________________________________

_______________________________________________________________

10. Nuclear power plants produce _____________________, which give off dangerous radiation.

11. What in the atmosphere has increased since the Industrial Revolution?

_______________________________________________________________

12. What do many scientists believe has happened as the amount of carbon dioxide in the atmosphere has increased?

_______________________________________________________________

_______________________________________________________________

_______________________________________________________________

13. What effect could the rise in global temperatures have on the world's oceans?

_______________________________________________________________

_______________________________________________________________

14. How is noise pollution harmful?

_______________________________________________________________

## RESOURCE DEPLETION

**Match the correct description with the correct term. Write the letter in the space provided.**

_____ 15. A resource that can be replaced at the same rate at which it is used is a(n)
- a. renewable resource.
- b. nonrenewable resource.
- c. mineral or fossil fuel.
- d. exhaustible resource.

_____ 16. A resource that cannot be replaced or is replaced over thousands or millions of years is a
- a. renewable resource.
- b. nonrenewable resource.
- c. wind energy resource.
- d. timber resource.

_____ 17. What is true about using oil or coal for energy?
- a. Coal and oil will last forever.
- b. Coal and oil may eventually run out.
- c. Coal and oil help the environment.
- d. Coal and oil will become easier to find.

_____ 18. What might happen in areas where fresh water is used faster than it is replaced?
a. Plants and animals will survive without water.
b. People might drink less water.
c. The areas might run out of fresh water.
d. There might be no change in the areas.

**EXOTIC SPECIES**

19. What is an exotic species?

_______________________________________________________________

_______________________________________________________________

20. Describe how exotic species might be moved from one part of the world to another.

_______________________________________________________________

_______________________________________________________________

**HUMAN POPULATION GROWTH**

_____ 21. What advances have made human population growth possible?
a. advances in housing and education
b. advances in radio and television
c. advances in medicine and farming
d. advances in farming and housing

_____ 22. What word describes the presence of too many individuals in an area for the available resources?
a. overpopulation
b. crowding
c. excessiveness
d. biodiversity

**HABITAT DESTRUCTION**

23. The place where an organism lives is its ____________________.

24. The number and variety of organisms in a given area during a specific period of time is ____________________.

25. What is the effect of habitat damage or destruction on biodiversity?

_______________________________________________________________

_______________________________________________________________

26. The clearing of forest lands is known as ____________________.

27. Describe one way deforestation is harmful to tropical rain forest habitats.

_______________________________________________________________

_______________________________________________________________

28. Pollution that comes from one source is called ______________________ pollution.

29. An oil spill is one example of ______________________ pollution in a marine habitat.

30. Pollution that comes from many sources is called ______________________ pollution.

31. Chemicals washed from land into rivers, lakes, and oceans are an example of ______________________ pollution in a marine environment.

32. How do plastics harm water habitats?

_______________________________________________________________

_______________________________________________________________

_______________________________________________________________

_______________________________________________________________

**EFFECTS ON HUMANS**

33. Describe how water pollution can harm humans.

_______________________________________________________________

_______________________________________________________________

34. Describe how air pollution can harm humans.

_______________________________________________________________

_______________________________________________________________

35. What is a possible long-term effect of exposure to some chemicals?

_______________________________________________________________

_______________________________________________________________

_______________________________________________________________

Name________________________________ Class __________________ Date __________________

Skills Worksheet

# Directed Reading A

## Section: Environmental Solutions

1. List four human needs that will have an impact on the Earth.

______________________________________________________________

______________________________________________________________

______________________________________________________________

______________________________________________________________

### CONSERVATION

_____ 2. The preservation and wise use of natural resources is called
- a. recycling.
- b. biodiversity.
- c. conservation.
- d. ecology.

_____ 3. Riding a bike not only saves fuel, it can also help prevent
- a. air pollution.
- b. water pollution.
- c. soil pollution.
- d. overpopulation.

_____ 4. When people practice conservation, they use fewer
- a. muscles.
- b. foods.
- c. natural resources.
- d. habitats.

_____ 5. Which of the following is NOT one of the three Rs of conservation?
- a. reuse
- b. recycle
- c. reduce
- d. rebuild

### REDUCE

6. In some countries, as much as much as one-third of the waste produced is made up of what?

______________________________________________________________

7. Something that is ______________________ can be broken down by living organisms.

8. Describe how some farmers are taking better care of the environment.

______________________________________________________________

______________________________________________________________

______________________________________________________________

______________________________________________________________

9. What are some new sources of energy being studied in the hopes that they will replace fossil fuels?

_______________________________________________

_______________________________________________

10. What have car companies developed that will reduce the need for fossil fuels and reduce pollution?

_______________________________________________

## REUSE

11. What is one way plastic bags can be reused?

_______________________________________________

_______________________________________________

12. What are two ways old tires can be reused?

_______________________________________________

_______________________________________________

13. How is water reclaimed and reused?

_______________________________________________

_______________________________________________

_______________________________________________

_______________________________________________

## RECYCLE

_____ 14. Recovering valuable or useful materials from waste or scrap is called

a. reusing.
b. reducing.
c. renewing.
d. recycling.

_____ 15. What can be recycled into a natural fertilizer?

a. old tires
b. old plastics
c. yard clippings
d. aluminum cans

_____ 16. What percentage of energy needed to change raw ore into aluminum is saved by recycling aluminum?

a. 95%
b. 75%
c. 16%
d. 100%

_____ 17. One example of resource recovery is using garbage to
- a. generate electricity.
- b. reduce air pollution.
- c. limit trash collection.
- d. save trees.

18. List three materials that can be recycled, and describe how one of them is used again.

______________________________________________________________________

______________________________________________________________________

______________________________________________________________________

19. How do some communities make recycling easy for citizens?

______________________________________________________________________

______________________________________________________________________

20. Why are some people concerned about burning waste material to produce energy?

______________________________________________________________________

______________________________________________________________________

______________________________________________________________________

**MAINTAINING BIODIVERSITY**

21. Why is biodiversity important to an ecosystem?

______________________________________________________________________

______________________________________________________________________

______________________________________________________________________

______________________________________________________________________

______________________________________________________________________

22. Describe what might happen if an important predator in an ecosystem were lost.

______________________________________________________________________

______________________________________________________________________

______________________________________________________________________

______________________________________________________________________

______________________________________________________________________

23. How does the Endangered Species Act help protect biodiversity?

_______________________________________________________________

_______________________________________________________________

_______________________________________________________________

24. What is one example of an animal that has been helped by the Endangered Species Act?

_______________________________________________________________

25. Why is it important to protect habitats and not just individual species of organisms?

_______________________________________________________________

_______________________________________________________________

**ENVIRONMENTAL STRATEGIES**

26. What does the EPA do to help protect the environment?

_______________________________________________________________

_______________________________________________________________

_______________________________________________________________

27. List five environmental strategies that people can use to help protect Earth's environment.

_______________________________________________________________

_______________________________________________________________

_______________________________________________________________

_______________________________________________________________

_______________________________________________________________

28. List five things that young people can do to help clean up Earth.

_______________________________________________________________

_______________________________________________________________

_______________________________________________________________

_______________________________________________________________

_______________________________________________________________

Name______________________________Class ________________ Date ________________

Skills Worksheet

# Vocabulary and Section Summary

## Environmental Problems

### VOCABULARY

**In your own words, write a definition of the following terms in the space provided.**

1. pollution

_______________________________________________

_______________________________________________

2. renewable resource

_______________________________________________

_______________________________________________

3. nonrenewable resource

_______________________________________________

_______________________________________________

4. overpopulation

_______________________________________________

_______________________________________________

5. biodiversity

_______________________________________________

_______________________________________________

### SECTION SUMMARY

**Read the following section summary.**

- Pollutants include garbage, chemicals, high-energy wastes, gases, and noise.
- Renewable resources can be used over and over. Nonrenewable resources cannot be replaced or are replaced over thousands or millions of years.
- Exotic species can become pests and compete with native species.
- Overpopulation happens when a population is so large that it can't get what it needs to survive.
- Habitat destruction can lead to soil erosion, water pollution, and decreased biodiversity.
- In addition to harming the environment, pollution can harm humans.

Name________________________________Class ___________________ Date ___________________

Skills Worksheet

# Vocabulary and Section Summary

## Environmental Solutions

### VOCABULARY

**In your own words, write a definition of the following terms in the space provided.**

1. conservation

_______________________________________________________________

_______________________________________________________________

2. recycling

_______________________________________________________________

_______________________________________________________________

### SECTION SUMMARY

**Read the following section summary.**

- Conservation is the preservation and wise use of natural resources. Conservation helps reduce pollution, ensures that resources will be available in the future, and protects habitats.
- The three Rs are Reduce, Reuse, and Recycle. Reducing means using fewer resources. Reusing means using materials and products over and over. Recycling is the recovery of materials from waste.
- Biodiversity is vital for maintaining healthy ecosystems. A loss of one species can affect an entire ecosystem.
- Biodiversity can be preserved by protecting endangered species and entire habitats.
- Environmental strategies include reducing pollution, reducing pesticide use, protecting habitats, enforcing the Endangered Species Act, and developing alternative energy resources.

Name______________________________Class ________________ Date ________________

Skills Worksheet

# Directed Reading A

## Section: What Is Matter?

### MATTER

1. What characteristic do a human, hot soup, the metal wires in a toaster, and the glowing gases in a neon sign have in common?

_____________________________________________________________________________

2. What is matter?

_____________________________________________________________________________

### MATTER AND VOLUME

_____ 3. What unit would you use to measure the amount of water in a lake?
a. grams (g)
b. liters (L)
c. meters (m)
d. milliliters (mL)

_____ 4. What unit would you use to measure the volume of soda in a can?
a. centimeters (cm)
b. grams (g)
c. liters (L)
d. milliliters (mL)

5. What is volume?

_____________________________________________________________________________

6. Things with ________________________ cannot share the same space at the same time.

7. To measure a volume of water in a graduated cylinder, you should look at the bottom of the curve at the surface of the water called the ________________________.

8. The volume of solid objects is commonly expressed in ________________________ units.

9. What three dimensions are needed to find the volume of a rectangular solid?

_____________________________________________________________________________

_____________________________________________________________________________

_____________________________________________________________________________

10. How could the volume of a 12-sided object be found using water and a graduated cylinder?

_____________________________________________________________________________

11. Why can you express the volume of the 12-sided object measured by this method in cubic units?

_______________________________________________________________

## MATTER AND MASS

_____ 12. The amount of matter in an object is its
  a. volume.
  b. length.
  c. meniscus.
  d. mass.

_____ 13. The SI unit of mass is the
  a. newton.
  b. liter.
  c. kilogram.
  d. pound.

_____ 14. The SI unit of weight is the
  a. newton.
  b. liter.
  c. kilogram.
  d. pound.

_____ 15. One newton is equal to the weight of an object that has
  a. a mass of 100 g on the moon.
  b. a volume of 1 $m^3$ on Earth.
  c. a mass of 1,000 g on Earth.
  d. a mass of 100 g on Earth.

16. What is the only way to change the mass of an object?

_______________________________________________________________

_______________________________________________________________

**For each description, write whether it applies to mass or to weight.**

______________________ 17. is always constant no matter where the object is located

______________________ 18. is a measure of the gravitational force on an object

______________________ 19. is measured using a spring scale

______________________ 20. is expressed in grams (g), kilograms (kg), or milligrams (mg)

______________________ 21. is expressed in newtons (N)

______________________ 22. is less on the moon than on Earth

______________________ 23. is a measure of the amount of matter in the object

## INERTIA

_____ 24. The tendency of an object to resist a change in motion is known as
a. mass.
b. gravitation.
c. inertia.
d. weight.

25. What is needed in order to cause an object at rest to move, or an object in motion to change its direction or speed?

_______________________________________________

_______________________________________________

26. How does mass affect the inertia of an object?

_______________________________________________

_______________________________________________

_______________________________________________

27. Why is it harder to get a cart full of potatoes moving than one that is empty?

_______________________________________________

_______________________________________________

_______________________________________________

_______________________________________________

Name_______________________________ Class ________________ Date ________________

Skills Worksheet

# Directed Reading A

## Section: Physical Properties

### PHYSICAL PROPERTIES

_____ 1. A characteristic of matter that can be observed or measured without changing the identity of the matter is a
a. matter property.
b. physical property.
c. chemical property.
d. volume property.

_____ 2. Some examples of physical properties are
a. color, odor, and age.
b. color, odor, and speed.
c. color, odor, and magnetism.
d. color, odor, and anger.

**Match the correct example with the correct physical property. Write the letter in the space provided.**

_____ 3. Aluminum can be flattened into sheets of foil.

_____ 4. An ice cube floats in a glass of water.

_____ 5. Copper can be pulled into thin wires.

_____ 6. Plastic foam protects you from hot liquid.

_____ 7. Flavored drink mix dissolves in water.

_____ 8. An onion gives off a very distinctive smell.

_____ 9. A golf ball has more mass than a table tennis ball.

a. state
b. solubility
c. thermal conductivity
d. malleability
e. odor
f. ductility
g. density

10. Density is the ______________________ that describes the relationship between mass and volume.

11. Objects such as a cotton ball and a small tomato can occupy similar volumes but vary greatly in ______________________.

12. If you pour different liquids into a graduated cylinder, the liquids will form layers based upon differences in the ______________________ of each liquid.

13. Which layer of liquid would settle on the bottom of a graduated cylinder?

_______________________________________________________________

14. Where will the least dense liquid be found?

______________________________________________________________

15. Why would 1 kg of lead be less awkward to carry around than 1 kg of feathers?

______________________________________________________________

______________________________________________________________

______________________________________________________________

16. What will happen to a solid object made from matter with a greater density than water when it is dropped into water?

______________________________________________________________

______________________________________________________________

______________________________________________________________

17. How will knowing the density of a substance help you determine whether an object made from that material will float in water?

______________________________________________________________

______________________________________________________________

______________________________________________________________

18. What is the equation for density?

______________________________________________________________

19. What do *D, V*, and *m* stand for in the equation for density?

______________________________________________________________

______________________________________________________________

______________________________________________________________

20. The units for density take the form of a mass unit divided by a(n)

________________________ unit.

21. What are two reasons why density is a useful property for identifying substances?

______________________________________________________________

______________________________________________________________

**PHYSICAL CHANGES DO NOT FORM NEW SUBSTANCES**

22. A change that affects only the physical properties of a substance is known as a(n) ______________________.
23. What kind of changes are melting and freezing?

_______________________________________________

**Identify which of the following activities represent physical changes by writing PC in the space provided if they cause only physical changes. Put an X beside any that do not.**

_____ 24. sanding a piece of wood

_____ 25. baking bread

_____ 26. crushing an aluminum can

_____ 27. melting an ice cube

_____ 28. dissolving sugar in water

_____ 29. molding a piece of silver

30. When a substance undergoes a physical change, its ____________________ does not change.
31. What is changed when matter undergoes a physical change? Give an example to explain your answer.

_______________________________________________

_______________________________________________

_______________________________________________

_______________________________________________

_______________________________________________

_______________________________________________

_______________________________________________

Name________________________________ Class __________________ Date __________________

Skills Worksheet

# Directed Reading A

## Section: Chemical Properties

### CHEMICAL PROPERTIES

_____ 1. The property of matter that describes its ability to change into new matter with different properties is known as a
a. chemical change. c. chemical property.
b. physical change. d. physical property.

_____ 2. The chemical property that describes the ability of two or more substancesto combine to form new substances is called
a. reactivity. c. density.
b. flammability. d. solubility.

_____ 3. The ability of a substance to burn is a chemical property known as
a. reactivity. c. density.
b. flammability. d. solubility.

_____ 4. An iron nail is reactive with
a. rubbing alcohol.
b. other iron nails.
c. wood in a house.
d. oxygen in the air.

_____ 5. Which of the following statements is true about characteristic properties of matter?
a. Characteristic properties depend on the size of the sample.
b. Characteristic properties may be either physical or chemical properties.
c. Characteristic properties involve only chemical properties.
d. Characteristic properties involve only the physical nature of the matter.

6. Describe the ways that burning changes the nature of wood.

______________________________________________________________

______________________________________________________________

7. A substance always has ________________________ properties, even though they are difficult to observe.

8. Scientists use ________________________ properties to help them identify and classify matter.

## CHEMICAL CHANGES AND NEW SUBSTANCES

_____ 9. Chemical changes are the processes by which substances
a. move from place to place.
b. change into new substances.
c. change in their physical properties.
d. become greater in mass.

_____ 10. Which of the following would NOT be considered an example of a chemical change?
a. the bubbling action of effervescent tablets
b. the green coating on copper statues
c. the melting of a Popsicle
d. the burning of rocket fuel

11. How do you know that baking a cake involves chemical changes?

______________________________________________________________

______________________________________________________________

______________________________________________________________

______________________________________________________________

12. List some signs or clues that show that a change you are observing is a chemical change.

______________________________________________________________

______________________________________________________________

______________________________________________________________

______________________________________________________________

13. Because ______________________ change the identity of the substances involved, they are hard to reverse.

14. How could some chemical changes be reversed? Give an example.

______________________________________________________________

______________________________________________________________

______________________________________________________________

______________________________________________________________

## PHYSICAL VERSUS CHEMICAL CHANGES

_____ 15. What is the most important question to ask to determine whether a change is physical or chemical?
a. Was there a color change?
b. Did the composition change?
c. Was there a change in size?
d. Did the change involve a change in state?

_____ 16. What is the name of the process by which water is broken down into hydrogen and oxygen using an electric current?
a. electrolysis
b. decomposition
c. reactivity
d. reversibility

17. During ________________________, the composition of a substance does not change.

**Identify whether the following changes are physical changes or chemical changes. Label each change either PC for physical change or CC for chemical change.**

_____ 18. mixing vinegar and baking soda

_____ 19. grinding baking soda into a powder

_____ 20.souring milk

_____ 21. melting an ice cream bar

_____ 22. burning a wooden match

_____ 23. shooting off fireworks

_____ 24. mixing drink mix into water

_____ 25. bending an iron nail

Name______________________________Class __________________ Date __________________

Skills Worksheet

# Vocabulary and Section Summary

## What Is Matter?

### VOCABULARY

**In your own words, write a definition of the following terms in the space provided.**

1. matter

_______________________________________________________________

_______________________________________________________________

2. volume

_______________________________________________________________

_______________________________________________________________

3. meniscus

_______________________________________________________________

_______________________________________________________________

4. mass

_______________________________________________________________

_______________________________________________________________

5. weight

_______________________________________________________________

_______________________________________________________________

6. inertia

_______________________________________________________________

_______________________________________________________________

## SECTION SUMMARY

**Read the following section summary.**

- Two properties of matter are volume and mass.
- Volume is the amount of space taken up by an object.
- The SI unit of volume is the liter (L).
- Mass is the amount of matter in an object.
- The SI unit of mass is the kilogram (kg).
- Weight is a measure of the gravitational force on an object, usually in relation to the Earth.
- Inertia is the tendency of an object to resist being moved or, if the object is moving, to resist a change in speed or direction. The more massive an object is, the greater its inertia.

Name_______________________________ Class _________________ Date _________________

Skills Worksheet

# Vocabulary and Section Summary

## Physical Properties

### VOCABULARY

**In your own words, write a definition of the following terms in the space provided.**

1. physical property

_______________________________________________________________

_______________________________________________________________

2. density

_______________________________________________________________

_______________________________________________________________

3. physical change

_______________________________________________________________

_______________________________________________________________

### SECTION SUMMARY

**Read the following section summary.**

- Physical properties of matter can be observed without changing the identity of the matter.
- Examples of physical properties are conductivity, state, malleability, ductility, solubility, and density.
- Density is the amount of matter in a given space.
- Density is used to identify substances because the density of a substance is always the same at a given pressure and temperature.
- When a substance undergoes a physical change, its identity stays the same.
- Examples of physical changes are freezing, cutting, bending, dissolving,and melting.

Name_________________ Class _____________ Date _____________

Skills Worksheet

# Vocabulary and Section Summary

## Chemical Properties

### VOCABULARY

**In your own words, write a definition of the following terms in the space provided.**

1. chemical property

_______________________________________________

_______________________________________________

2. chemical change

_______________________________________________

_______________________________________________

### SECTION SUMMARY

**Read the following section summary.**

- Chemical properties describe a substance based on its ability to change into a new substance that has different properties.
- Chemical properties can be observed only when a chemical change might happen.
- Examples of chemical properties are flammability and reactivity.
- New substances form as a result of a chemical change.
- Unlike a chemical change, a physical change does not alter the identity of a substance.

Name______________________ Class ______________ Date ______________

Skills Worksheet

# Directed Reading A

## Section: Three States of Matter

1. What are the three most familiar states of matter?

______________________________________________

______________________________________________

______________________________________________

2. What is a state of matter?

______________________________________________

______________________________________________

### PARTICLES OF MATTER

3. Matter is made up of ______________________ and ______________________.

**Match the correct description with the correct state of matter. Write the letter in the space provided.**

a. solid
b. liquid
c. gas

_____ 4. Particles do not move fast enough to overcome the strong attraction between them.

_____ 5. Particles move independently of each other.

_____ 6. Particles are close together but can slide past one another.

_____ 7. Particles are close together and vibrate in place.

_____ 8. Particles move fast enough to overcome nearly all of the attraction between them.

### SOLIDS

_____ 9. The particles of matter that make up a solid
a. have a weaker attraction than those of a liquid.
b. do not move at all.
c. do not move fast enough to overcome the force of attraction.
d. move from place to place.

10. What is a solid?

______________________________________________________________________

______________________________________________________________________

11. How are the particles in a crystalline solid arranged?

______________________________________________________________________

______________________________________________________________________

12. How are the particles in an amorphous solid arranged?

______________________________________________________________________

______________________________________________________________________

**LIQUIDS**

13. How do the particles of a liquid make it possible to pour juice into a glass?

______________________________________________________________________

______________________________________________________________________

14. A beaker and a cylinder each contain 350 mL of juice. What does this show you about the properties of a liquid?

______________________________________________________________________

______________________________________________________________________

15. Liquids tend to form spherical droplets because of ________________________ tension.

16. Water has a lower ________________________ than honey.

**GASES**

17. What is a gas?

______________________________________________________________________

______________________________________________________________________

18. How is it possible for one tank of helium to fill 700 balloons?

______________________________________________________________________

______________________________________________________________________

______________________________________________________________________

Name________________________________ Class __________________ Date __________________

Skills Worksheet

# Directed Reading A

## Section: Behavior of Gases

### DESCRIBING GAS BEHAVIOR

_____ 1. What state of matter is helium?
- a. solid
- b. liquid
- c. gas
- d. plasma

2. A measure of how fast the particles in an object are moving is the

________________________.

3. Why is more gas needed to fill helium balloons on a cold day?

______________________________________________________________

______________________________________________________________

______________________________________________________________

4. The amount of space that an object takes up is the ________________________.

5. The volume of any gas depends upon the size of the

________________________.

6. The amount of force exerted on a given area is called

________________________.

7. Why does the basketball have greater pressure than the beachball?

______________________________________________________________

______________________________________________________________

______________________________________________________________

______________________________________________________________

### GAS BEHAVIOR LAWS

_____ 8. Lifting a piston on a cylinder of gas shows that when the pressure of the gas
- a. increases, the temperature increases.
- b. decreases, the volume increases.
- c. decreases, the volume decreases.
- d. increases, the volume increases.

_____ 9. All of the following remain constant for Charles's law EXCEPT
a. the type of piston.
b. the amount of gas.
c. the volume of the gas.
d. the pressure.

10. The relationship between the volume and pressure of a gas is called

_________________________.

11. Weather balloons are only partially inflated before they're released into the atmosphere. Why is that?

_______________________________________________________________

_______________________________________________________________

_______________________________________________________________

_______________________________________________________________

12. Putting a balloon in the freezer is one way to demonstrate

_________________________.

13. The relationship between the volume and the temperature of a gas when pressure remains constant is known as _________________________.

Name______________________________ Class ________________ Date ________________

Skills Worksheet

# Directed Reading A

## Section: Changes of State

### ENERGY AND CHANGES OF STATE

_____ 1. Which has the most energy?
a. particles in steam
b. particles in liquid water
c. particles in ice
d. particles in freezing water

2. When a substance changes from one physical form to another, we say the substance has had a(n) ______________________.

3. List the five changes of state.

______________________________________________________________

______________________________________________________________

______________________________________________________________

______________________________________________________________

______________________________________________________________

### MELTING: SOLID TO LIQUID

4. Could you use gallium to make jewelry? Why or why not?

______________________________________________________________

______________________________________________________________

______________________________________________________________

______________________________________________________________

5. The temperature at which a substance changes from solid to liquid is the ______________________ of the substance.

6. Melting is considered a(n) ______________________ change because energy is gained by the substance as it changes state.

### FREEZING: LIQUID TO SOLID

7. A substance's ______________________ is the temperature at which it changes from a liquid to a solid.

8. What happens if energy is added or removed from a glass of ice water?

_______________________________________________

_______________________________________________

_______________________________________________

_______________________________________________

9. Freezing is considered a(n) ____________________ change because energy is removed from the substance.

## EVAPORATION: LIQUID TO GAS

**Match the correct definition with the correct term. Write the letter in the space provided.**

_____ 10. the change of a substance from a liquid to a gas

_____ 11. the change of state from a liquid to a gas when the vapor pressure equals the atmospheric pressure

_____ 12. the pressure inside the bubbles of a boiling liquid

_____ 13. the temperature at which a liquid boils

a. boiling point
b. vapor pressure
c. evaporation
d. boiling

14. As you go higher above sea level, the ____________________ decreases and the ____________________ of a substance gets lower.

## CONDENSATION: GAS TO LIQUID

15. The change of state from a gas to a liquid is ____________________.

16. At a given pressure, the condensation point for a substance is the same as its ____________________.

17. For a substance to change from a gas to a liquid, particles must ____________________.

## SUBLIMATION: SOLID TO GAS

18. Solid carbon dioxide isn't ice. So why is it called "dry ice"?

______________________________________________________________________

______________________________________________________________________

______________________________________________________________________

19. The change of state from a solid to a gas is called ___________________________.

## CHANGE OF TEMPERATURE VS. CHANGE OF STATE

20. The speed of the particles in a substance changes when the

    ___________________________ changes.

21. The temperature of a substance does not change before the

    ___________________________ is complete.

Name________________________________Class ___________________ Date __________________

Skills Worksheet

# Vocabulary and Section Summary

## Three States of Matter

### VOCABULARY

**In your own words, write a definition of the following terms in the space provided.**

1. states of matter

_______________________________________________________________________________

_______________________________________________________________________________

2. solid

_______________________________________________________________________________

_______________________________________________________________________________

3. liquid

_______________________________________________________________________________

_______________________________________________________________________________

4. surface tension

_______________________________________________________________________________

_______________________________________________________________________________

5. viscosity

_______________________________________________________________________________

_______________________________________________________________________________

6. gas

_______________________________________________________________________________

_______________________________________________________________________________

### SECTION SUMMARY

**Read the following section summary.**

- The three most familiar states of matter are solid, liquid, and gas.
- All matter is made of tiny particles called atoms and molecules that attract each other and move constantly.
- A solid has a definite shape and volume.
- A liquid has a definite volume but not a definite shape.
- A gas does not have a definite shape or volume.

Name________________________________Class ________________ Date ________________

Skills Worksheet

# Vocabulary and Section Summary

## Behavior of Gases

### VOCABULARY

**In your own words, write a definition of the following terms in the space provided.**

1. temperature

_______________________________________________

_______________________________________________

2. volume

_______________________________________________

_______________________________________________

3. pressure

_______________________________________________

_______________________________________________

4. Boyle's law

_______________________________________________

_______________________________________________

5. Charles's law

_______________________________________________

_______________________________________________

### SECTION SUMMARY

**Read the following section summary.**

- Temperature measures how fast the particles in an object are moving.
- Gas pressure increases as the number of collisions of gas particles increases.
- Boyle's law states that if the temperature doesn't change, the volume of a gas increases as the pressure decreases.
- Charles's law states that if the temperature doesn't change, the volume of a gas increases.

Name_______________ Class __________ Date __________

Skills Worksheet

# Vocabulary and Section Summary

## Changes of State

### VOCABULARY

**In your own words, write a definition of the following terms in the space provided.**

1. change of state

______________________________

______________________________

2. melting

______________________________

______________________________

3. evaporation

______________________________

______________________________

4. boiling

______________________________

______________________________

5. condensation

______________________________

______________________________

6. sublimation

______________________________

______________________________

## SECTION SUMMARY

**Read the following section summary.**

- A change of state is the conversion of a substance from one physical form to another.
- Energy is added during endothermic changes. Energy is removed during exothermic changes.
- The freezing point and the melting point of a substance are the same temperature.
- Both boiling and evaporation result in a liquid changing to a gas.
- Condensation is the change of a gas to a liquid. It is the reverse of evaporation.
- Sublimation changes a solid directly to a gas.
- The temperature of a substance does not change during a change of state.

Name_______________________________ Class ___________________ Date ___________________

Skills Worksheet

# Directed Reading A

## Section: Elements

_____ 1. Which of the following is NOT a physical or chemical change?
a. crushing
b. weighing
c. melting
d. passing electric current

### ELEMENTS, THE SIMPLEST SUBSTANCES

2. A pure substance that cannot be separated into simpler substances by physical or chemical means is a(n) ________________________.

3. A substance that contains only one type of particle is a(n) ________________________.

### PROPERTIES OF ELEMENTS

4. The amount of an element present does not affect the element's ________________________.

5. Why does a helium-filled balloon float up when it is released?

_______________________________________________________________

_______________________________________________________________

**Look at each property listed below. If it is a characteristic property of elements, write CP on the line. If it is not a characteristic property, write N.**

_____ 6. size

_____ 7. melting point

_____ 8. density

_____ 9. shape

_____ 10. mass

_____ 11. volume

_____ 12. color

_____ 13. hardness

_____ 14. flammability

_____ 15. weight

_____ 16. reactivity with acid

## CLASSIFYING ELEMENTS BY THEIR PROPERTIES

17. What are two common properties that most terriers share?

_______________________________________________

_______________________________________________

18. All elements can be classified as metals, metalloids, or

_______________________.

19. An element that is shiny and that conducts heat and electric current well is

a(n) _______________________.

20. An element that conducts heat and electric current poorly, and can be a solid,

liquid, or gas is a(n) _______________________.

21. Elements that have properties of both metals and nonmetals are

_______________________.

**Indicate whether the description applies to a metal, a nonmetal, or a metalloid. Write the correct letter in the space provided.**

a. metalloids
b. nonmetals
c. metals

_____ 22. are malleable

_____ 23. are dull or shiny

_____ 24. are poor conductors

_____ 25. tend to be brittle and unmalleable as solids

_____ 26. are always shiny

_____ 27. are also called semiconductors

_____ 28. are always dull

_____ 29. are somewhat ductile

_____ 30. include boron, silicon, antimony

_____ 31. include lead, tin, copper

_____ 32. include sulfur, iodine, neon

Name______________________________ Class _______________ Date _______________

Skills Worksheet

# Directed Reading A

## Section: Compounds

1. List three examples of compounds you encounter every day.

_______________________________________________

_______________________________________________

### COMPOUNDS: MADE OF ELEMENTS

_____ 2. Which of the following is NOT true about compounds?
   a. Compounds are combinations of elements that join in specific ratios according to their masses.
   b. The mass ratio of a specific compound is always the same.
   c. Compounds are random combinations of elements.
   d. Different mass ratios mean different compounds.

3. When two or more elements are joined by chemical bonds to form a new pure substance, we call that new substance a(n) ____________________.

4. A compound is different from the ____________________ that reacted to form it.

### PROPERTIES OF COMPOUNDS

_____ 5. Which of the following statements is true about the properties of compounds?
   a. A property of all compounds is to react with acid.
   b. Each compound has its own physical properties.
   c. Compounds cannot be identified by their chemical properties.
   d. A compound has the same properties as the elements that form it.

6. Sodium and chlorine can be extremely dangerous in their elemental form. How is it possible that we can eat them in a compound?

_______________________________________________

_______________________________________________

_______________________________________________

**Match the correct description with the correct term. Write the letter in the space provided.**

_____ 7. a poisonous, greenish yellow gas

_____ 8. table salt

_____ 9. a soft, silvery white metal that reacts violently with water

a. sodium chloride
b. chlorine
c. sodium

**BREAKING DOWN COMPOUNDS**

10. What compound helps give carbonated beverages their "fizz"?

_______________________________________________________________

_______________________________________________________________

11. Which elements make up the compound that helps give carbonated beverages their "fizz"?

_______________________________________________________________

_______________________________________________________________

12. The only way to break down a compound is through a(n)

_______________________ change.

**COMPOUNDS IN YOUR WORLD**

13. Aluminum is produced by breaking down the compound

_______________________.

14. Plants use the compound _______________________ in photosynthesis to make carbohydrates.

Name________________________________ Class ________________ Date ________________

Skills Worksheet

# Directed Reading A

## Section: Mixtures

1. A pizza is a(n) ______________________.

### PROPERTIES OF MIXTURES

2. A combination of two or more substances that are not chemically combined is a(n)______________________.
3. When two or more materials combine chemically, they form a(n) ______________________.
4. How can you tell that a pizza is a mixture?

______________________________________________________________________

______________________________________________________________________

5. Mixtures are separated through ______________________ changes.

**Match the correct method of separation with the each substance. Write the letter in the space provided. Each method may be used only once.**

_____ 6. crude oil

_____ 7. a mixture of aluminum and iron

_____ 8. parts of blood

_____ 9. sulfur and salt

a. distillation
b. magnet
c. filter
d. centrifuge

10. Granite can be pink or black, depending on the ______________________ of feldspar, mica, and quartz.

### SOLUTIONS

_____ 11. Which of the following is NOT true of solutions?
a. They contain a dissolved substance called a solute.
b. They are composed of two or more evenly distributed substances.
c. They contain a substance called a solvent, in which another substance is dissolved.
d. They appear to be more than one substance.

12. The process in which particles of substances separate and spread evenly through a mixture is known as ______________________.

13. In a solution, the ________________________ is the substance that is dissolved,

    and the ________________________ is the substance in which it is dissolved.

14. Salt is ________________________ in water because it dissolves in water.

15. When two gases or two liquids form a solution, the substance that is present in

    the largest amount is the ________________________.

16. A solid solution of metals or nonmetals dissolved in metals is a(n)

    ________________________.

17. What can particles in solution NOT do because they are so small?

    ________________________________________________________________

    ________________________________________________________________

**CONCENTRATION OF SOLUTIONS**

18. A measure of the amount of solute dissolved in a solvent is called

    ________________________.

19. What is the difference between a dilute solution and a concentrated solution?

    ________________________________________________________________

    ________________________________________________________________

20. The ability of a solute to dissolve in a solvent at a certain temperature and

    pressure is called ________________________.

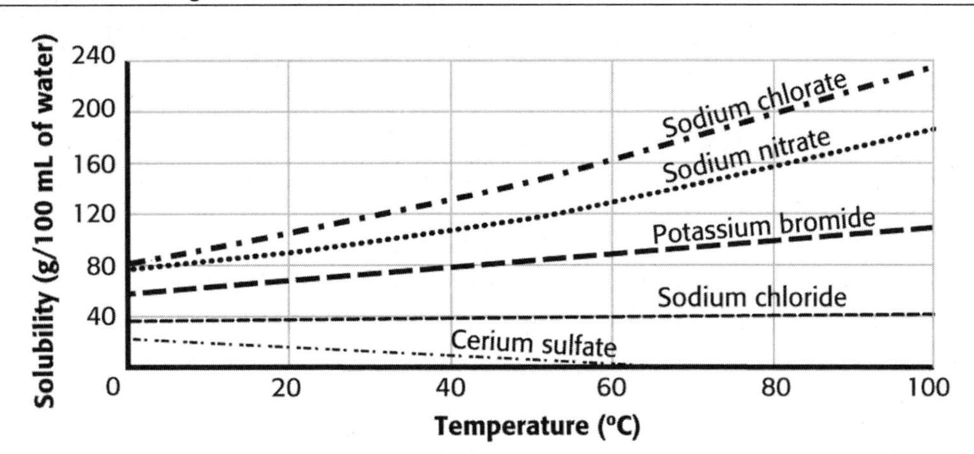


_____ 21. Look at the graph. Which solid is less soluble at higher temperatures than at lower temperatures?

a. sodium chloride
b. sodium nitrate
c. potassium bromide
d. cerium sulfate

_____ 22. Look at the graph. Which compound's solubility is least affected by temperature changes?

a. sodium chloride
b. sodium nitrate
c. potassium bromide
d. cerium sulfate

23. Solubility of solids in liquids tends to ____________________ with an increase in temperature.

24. Solubility of gases in liquids tends to ____________________ with an increase in temperature.

25. What are three ways to make a sugar cube dissolve more quickly in water?

_______________________________________________

_______________________________________________

_______________________________________________

**SUSPENSIONS**

_____ 26. Which of the following does NOT describe a suspension?
   a. Particles are soluble.
   b. Particles settle out over time.
   c. Particles can block light.
   d. Particles scatter light.

27. Why are the particles in a snow globe considered a suspension?

_______________________________________________

_______________________________________________

**COLLOIDS**

28. What do gelatin, milk, and stick deodorant have in common?

_______________________________________________

_______________________________________________

**Match the correct description with the correct term. Write the letter in the space provided.**

_____ 29. a mixture of two or more uniformly dispersed substances

_____ 30. a mixture in which particles of a material are more or less evenly dispersed throughout a liquid or gas

_____ 31. a mixture of particles that are large enough to scatter light but are not heavy enough to settle out

a. colloid
b. solution
c. suspension

Name______________________________Class ________________ Date ________________

Skills Worksheet

# Vocabulary and Section Summary

## Elements

### VOCABULARY

**In your own words, write a definition of the following terms in the space provided.**

1. element

______________________________________________________________

______________________________________________________________

2. pure substance

______________________________________________________________

______________________________________________________________

3. metal

______________________________________________________________

______________________________________________________________

4. nonmetal

______________________________________________________________

______________________________________________________________

5. metalloid

______________________________________________________________

______________________________________________________________

### SECTION SUMMARY

**Read the following section summary.**

- A substance in which all of the particles are alike is a pure substance.
- An element is a pure substance that cannot be broken down into anything simpler by physical or chemical means.
- Each element has a unique set of physical and chemical properties.
- Elements are classified as metals, nonmetals, or metalloids, based on their properties.

Name_____________________________Class _________________ Date _________________

Skills Worksheet

# Vocabulary and Section Summary

## Compounds

### VOCABULARY

**In your own words, write a definition of the following term in the space provided.**

1. compound

_______________________________________________________________________

_______________________________________________________________________

### SECTION SUMMARY

**Read the following section summary.**

- A compound is a pure substance composed of two or more elements.
- The elements that form a compound always combine in a specific ratio according to their masses.
- Each compound has a unique set of physical and chemical properties that differ from those of the elements that make up the compound.
- Compounds can be broken down into simpler substances only by chemical changes.

Name________________________________Class ____________________ Date ____________________

Skills Worksheet

# Vocabulary and Section Summary

## Mixtures

### VOCABULARY

**In your own words, write a definition of the following terms in the space provided.**

1. mixture

______________________________________________________________

______________________________________________________________

2. solution

______________________________________________________________

______________________________________________________________

3. solute

______________________________________________________________

______________________________________________________________

4. solvent

______________________________________________________________

______________________________________________________________

5. concentration

______________________________________________________________

______________________________________________________________

6. solubility

______________________________________________________________

______________________________________________________________

7. suspension

______________________________________________________________

______________________________________________________________

8. colloid

______________________________________________________________

______________________________________________________________

## SECTION SUMMARY

**Read the following section summary.**

- A mixture is a combination of two or more substances, each of which keeps its own characteristics.
- Mixtures can be separated by physical means, such as filtration and evaporation.
- A solution is a mixture that appears to be a single substance but is composed of a solute dissolved in a solvent.
- Concentration is a measure of the amount of solute dissolved in a solvent.
- The solubility of a solute is the ability of the solute to dissolve in a solvent at a certain temperature.
- Suspensions are mixtures that contain particles large enough to settle out or be filtered and to block or scatter light.
- Colloids are mixtures that contain particles that are too small to settle out or be filtered but are large enough to scatter light.

Name________________________________ Class ________________ Date ________________

Skills Worksheet

# Directed Reading A

## Section: Development of the Atomic Theory

### THE BEGINNING OF ATOMIC THEORY

_____ 1. The word *atom* comes from the Greek word *atomos,* which means
a. "dividable."
b. "invisible."
c. "hard particles."
d. "not able to be divided."

_____ 2. Which of the following statements is a part of Democritus's theory about atoms?
a. Atoms are small, soft particles.
b. Atoms are always standing still.
c. Atoms are made of a single material.
d. Atoms are small particles that can be cut in half again and again.

3. We know that Democritus was right to say that all matter was made up of atoms. So why did people ignore Democritus's ideas for such a long time?

_______________________________________________________________

_______________________________________________________________

_______________________________________________________________

4. The smallest unit of an element that maintains the properties of that element is

a(n) ______________________.

### DALTON'S ATOMIC THEORY BASED ON EXPERIMENTS

_____ 5. Which of the following was NOT one of Dalton's theories?
a. All substances are made of atoms.
b. Atoms of the same element are exactly alike.
c. Atoms of different elements are alike.
d. Atoms join with other atoms to make new substances.

6. Dalton experimented with different substances. What did his results suggest?

_______________________________________________________________

_______________________________________________________________

_______________________________________________________________

**THOMSON'S DISCOVERY OF ELECTRONS**

7. In Thomson's experiments with a cathode-ray tube, he discovered that a(n) ______________________ charged plate attracted the beam. He concluded that the beam was made up of particles that have ______________________ electric charges.

8. The negatively charged subatomic particles that Thomson discovered are now called ______________________.

9. In Thomson's "plum-pudding" model, electrons are mixed throughout an ______________________.

**RUTHERFORD'S ATOMIC "SHOOTING GALLERY"**

_____ 10. Before his experiment, what did Rutherford expect the particles to do?
a. He expected the particles to pass right through the gold foil.
b. He expected the particles to deflect to the sides of the gold foil.
c. He expected the particles to bounce straight back.
d. He expected the particles to become negatively charged.

11. What were the surprising results of Rutherford's gold-foil experiment?

______________________________________________________________

______________________________________________________________

______________________________________________________________

**WHERE ARE THE ELECTRONS?**

_____ 12. In 1911, Rutherford revised the atomic theory. Which of the following is NOT part of that theory?
a. Most of the atom's mass is in its nucleus.
b. The nucleus is a tiny, dense, positively charged region.
c. Positively charged particles that pass close by the nucleus are pushed away by the positive charges in the nucleus.
d. The nucleus is made up of protons and electrons.

13. The center of an atom is a dense region consisting of protons and neutrons called the ______________________.

14. What are electron clouds?

______________________________________________________________

Name________________________________Class __________________ Date __________________

Skills Worksheet

# Directed Reading A

## Section: The Atom

### HOW SMALL IS AN ATOM?

_____ 1. Which of the following statements is true?
a. A penny has about 20,000 atoms.
b. A penny has more atoms than Earth has people.
c. Aluminum is made up of large-sized atoms.
d. Aluminum atoms have a diameter of about 3 cm.

### WHAT IS AN ATOM MADE OF?

**Match the correct description with the correct term. Write the letter in the space provided.**

_____ 2. particle of the nucleus that has no electrical charge

_____ 3. particle found in the nucleus that is positively charged

_____ 4. particle with an unequal number of protons and electrons

_____ 5. negatively charged particle found outside the nucleus

_____ 6. contains most of the mass of an atom

_____ 7. SI unit used to express the masses of atomic particles

a. electron
b. atomic mass unit (amu)
c. nucleus
d. proton
e. ion
f. neutron

### HOW DO ATOMS OF DIFFERENT ELEMENTS DIFFER?

8. The simplest atom is the ______________________ atom. It has one ______________________ and one ______________________.

9. Neutrons in the atom's ______________________ keep two or more protons from moving apart.

10. If you build an atom using two protons, two neutrons, and two electrons, you have built an atom of ______________________.

11. An atom does not have to have equal numbers of ______________________ and ______________________.

12. The number of protons in the nucleus of an atom is the

    ______________________ of that atom.

## ISOTOPES

_____ 13. Isotopes always have
- a. the same number of protons.
- b. the same number of neutrons.
- c. a different atomic number.
- d. the same mass.

_____ 14. Which of the following is NOT true about unstable atoms?
- a. They are radioactive.
- b. They have a nucleus that always remains the same.
- c. They give off energy as they fall apart.
- d. They give off smaller particles as they fall apart.

_____ 15. What is the mass number of an isotope that has 5 protons, 6 neutrons, and 5 electrons?
- a. 1
- b. 11
- c. 10
- d. 16

_____ 16. If carbon has an atomic number of 6, how many neutrons does carbon-12 have?
- a. 12
- b. 8
- c. 6
- d. 18

17. Most elements contain a mixture of two or more ______________________.

18. The weighted average of the masses of all the naturally occurring isotopes of an element is the ______________________.

## FORCES IN ATOMS

**Match the correct definition with the correct term. Write the letter in the space provided.**

_____ 19. helps protons stay together in the nucleus

_____ 20. pulls objects toward one another

_____ 21. an important force in radioactive atoms

_____ 22. holds the electrons around the nucleus

- a. gravitational force
- b. electromagnetic force
- c. strong force
- d. weak force

Name________________________________ Class ____________________ Date ____________________

Skills Worksheet

# Vocabulary and Section Summary

## Development of the Atomic Theory

### VOCABULARY

**In your own words, write a definition of the following terms in the space provided.**

1. atom

_______________________________________________________________

_______________________________________________________________

2. electron

_______________________________________________________________

_______________________________________________________________

3. nucleus

_______________________________________________________________

_______________________________________________________________

4. electron cloud

_______________________________________________________________

_______________________________________________________________

### SECTION SUMMARY

**Read the following section summary.**

- Democritus thought that matter is composed of atoms.
- Dalton based his theory on observations of how elements combine.
- Thomson discovered electrons in atoms.
- Rutherford discovered that atoms are mostly empty space with a dense, positive nucleus.
- Bohr proposed that electrons are located in levels at certain distances from the nucleus.
- The electron-cloud model represents the current atomic theory.

Skills Worksheet

# Vocabulary and Section Summary

## The Atom

### VOCABULARY

**In your own words, write a definition of the following terms in the space provided.**

1. proton

___

___

2. atomic mass unit

___

___

3. neutron

___

___

4. atomic number

___

___

5. isotope

___

___

6. mass number

___

___

7. atomic mass

___

___

## SECTION SUMMARY

**Read the following section summary.**

- Atoms are extremely small. Ordinary-sized objects are made up of very large numbers of atoms.
- Atoms consist of a nucleus, which has protons and usually neutrons, and electrons, located in electron clouds around the nucleus.
- The number of protons in the nucleus of an atom is that atom's atomic number. All atoms of an element have the same atomic number.
- Different isotopes of an element have different numbers of neutrons in their nuclei. Isotopes of an element share most chemical and physical properties.
- The mass number of an atom is the sum of the atom's neutrons and protons.
- Atomic mass is a weighted average of the masses of natural isotopes of an element.
- The forces at work in an atom are gravitational force, electromagnetic force, strong force, and weak force.

Skills Worksheet

# Directed Reading A

## Section: Arranging the Elements

1. Why do you think scientists might have been frustrated by the organization of the elements before 1869?

______________________________________________________________________

______________________________________________________________________

______________________________________________________________________

______________________________________________________________________

______________________________________________________________________

______________________________________________________________________

### DISCOVERING A PATTERN

_____ 2. Which arrangement of elements did Mendeleev find produced a repeating pattern of properties?
a. by increasing density
b. by increasing melting point
c. by increasing shine
d. by increasing atomic mass

3. When something occurs or repeats at regular intervals, it is called

______________________.

4. Mendeleev's table, which shows elements' properties following a pattern that epeats every seven elements, is called the ______________________.

5. How was it possible that Mendeleev was able to predict the properties of elements that no one knew about?

______________________________________________________________________

______________________________________________________________________

______________________________________________________________________

______________________________________________________________________

______________________________________________________________________

______________________________________________________________________

## CHANGING THE ARRANGEMENT

_____ 6. How did Moseley solve the problem of the elements that did not fit the pattern according to their properties?
a. He rearranged the elements by atomic mass.
b. He discovered protons, neutrons, and electrons.
c. He disproved the periodic law.
d. He determined the elements' atomic number and then arranged them by atomic number.

7. When the repeating chemical and physical properties of elements change periodically with the elements' atomic numbers, it is called the

_________________________.

## PERIODIC TABLE OF THE ELEMENTS

_____ 3. Which information is NOT included in each square of the periodic table in your text?
a. atomic number
b. chemical symbol
c. melting point
d. atomic mass

9. How can you tell on the periodic table that carbon is a solid at room temperature?

_______________________________________________________________

_______________________________________________________________

_______________________________________________________________

## THE PERIODIC TABLE AND CLASSES OF ELEMENTS

10. Elements are classified as metals, nonmetals, or metalloids according to their

_________________________.

11. The number of _________________________ in the outer energy level of an atom helps determine which category an element belongs in.

12. How can the zigzag line on the periodic table help you?

_______________________________________________________________

_______________________________________________________________

_______________________________________________________________

13. Most elements are ______________________, which can be found to the left zigzag line on the periodic table.

14. Most metals are ______________________, which means that they can be drawn into thin wires.

15. Most metals are ______________________ at room temperature

16. Most metals are malleable. What does this mean?

_______________________________________________________________

_______________________________________________________________

17. What metal is flattened into sheets that are made into cans and foil?

_______________________________________________________________

18. What elements are found to the right of the zigzag line on the periodic table?

_______________________________________________________________

19. Semiconductors, also called ______________________, are the elements that border the zigzag line on the periodic table.

**DECODING THE PERIODIC TABLE**

_____ 20. Which elements often share properties?
a. those in a period
b. those in a group
c. those with the same color
d. those in a horizontal row

_____ 21. The physical and chemical properties of the elements change
a. within a group.
b. within a family.
c. across each period.
d. across each group.

22. For most elements, the ______________________ has one or two letters, with the first letter always capitalized.

23. Horizontal rows of elements on the periodic table are called ______________________.

24. Vertical columns of elements on the periodic table are called ______________________, or ______________________.

25. Some elements, such as ______________________, are named after scientists. Others, such as ______________________, are named after places.

Name______________________________ Class ________________ Date ________________

Skills Worksheet

# Directed Reading A

## Section: Grouping the Elements

_____ 1. What gives elements in a family or group similar properties?
a. the same atomic mass
b. the same number of protons in their nuclei
c. the same number of electrons in their outer energy level
d. the same number of total electrons

### GROUP 1: ALKALI METALS

_____ 2. Which of the following is NOT true of alkali metals?
a. They can be cut with a knife.
b. They are usually stored in water.
c. They are the most reactive of all the metals.
d. They can easily give away their outer electron.

3. Metals that share both physical and chemical properties are called

_________________________.

### GROUP 2: ALKALINE-EARTH METALS

4. Atoms of _________________________ have two outer-level electrons.

5. What are two products made from calcium compounds?

_______________________________________________________________

_______________________________________________________________

_______________________________________________________________

6. In what way does calcium help you?

_______________________________________________________________

_______________________________________________________________

_______________________________________________________________

7. Name three alkaline-earth metals besides calcium.

_______________________________________________________________

_______________________________________________________________

_______________________________________________________________

## GROUPS 3–12: TRANSITION METALS

_____ 8. Which of the following characteristics does NOT describe transition metals?
a. They are good conductors of thermal energy.
b. They are more reactive than alkali and alkaline-earth metals.
c. They have one or two electrons in the outer energy level.
d. They are denser than elements in Groups 1 and 2.

9. Metals that are less reactive than alkali metals and alkaline-earth metals are called ______________________.

10. How is mercury different from other transition metals?

________________________________________________________________

________________________________________________________________

11. Two rows of transition metals are placed at the bottom of the periodic table to save space. Elements in the first row are called ______________________. Elements in the second row are called ______________________.

12. Which lanthanide forms a compound that enables you to see red on a computer screen?

________________________________________________________________

13. Which actinide is used in some smoke detectors?

________________________________________________________________

## GROUP 13: BORON GROUP

14. Why did Emperor Napoleon III of France use aluminum dinnerware?

________________________________________________________________

________________________________________________________________

________________________________________________________________

15. What are some of the uses of aluminum?

________________________________________________________________

________________________________________________________________

________________________________________________________________

## GROUP 14: CARBON GROUP

16. The metalloids ______________________ and ______________________, both in Group 14, are used to make computer chips.
17. What are three compounds of carbon that are necessary for living things on Earth?

    ________________________________________________________________

18. The hardest material known is ______________________.
19. What are some of the uses of diamond?

    ________________________________________________________________

    ________________________________________________________________

20. What form of carbon is used as a pigment?

    ________________________________________________________________

## GROUP 15: NITROGEN GROUP

21. Nitrogen is a ______________________ at room temperature.
22. Each element in the Nitrogen Group has ______________________ electrons in the outer level.
23. Nitrogen from the air can react with what element to make ammonia for fertilizer?

    ________________________________________________________________

## GROUP 16: OXYGEN GROUP

24. How is oxygen different from the other four elements in Group 16?

    ________________________________________________________________

    ________________________________________________________________

25. The element ______________________ can be found as a yellow solid in nature and is used to make sulfuric acid.
26. Why is oxygen important?

    ________________________________________________________________

    ________________________________________________________________

    ________________________________________________________________

## GROUP 17: HALOGENS

27. The atoms of ________________________ need to gain only one electron to have a complete outer level.

28. What important use do the halogens iodine and chlorine have in common?

_______________________________________________________________________

_______________________________________________________________________

29. Halogens combine with most metals to form ________________________, such as ________________________.

30. How does chlorinating water help protect people?

_______________________________________________________________________

_______________________________________________________________________

_______________________________________________________________________

## GROUP 18: NOBLE GASES

_____ 31. Which of the following statements about noble gases is NOT true?
- a. They are colorless and odorless at room temperature.
- b. They have a complete set of electrons in their outer energy level.
- c. They normally react with other elements.
- d. All of them are found in Earth's atmosphere in small amounts.

32. The atoms of ________________________ have a full set of electrons in their outer level.

33. The low ________________________ of helium makes blimps and weather balloons float.

## HYDROGEN

_____ 34. Which of the following statements about hydrogen is NOT true?
- a. It is useful as rocket fuel.
- b. It is the most abundant element in the universe.
- c. Its physical properties are closer to those of nonmetals than to those of metals.
- d. It has two electrons in its outer energy level.

Name________________________________ Class ________________ Date ________________

Skills Worksheet

# Vocabulary and Section Summary

## Arranging the Elements

### VOCABULARY

**In your own words, write a definition of the following terms in the space provided.**

1. periodic

______________________________________________________________________

______________________________________________________________________

2. periodic law

______________________________________________________________________

______________________________________________________________________

3. period

______________________________________________________________________

______________________________________________________________________

4. group

______________________________________________________________________

______________________________________________________________________

### SECTION SUMMARY

**Read the following section summary**

- Mendeleev developed the first periodic table by listing the elements in order of increasing atomic mass. He used his table to predict that elements with certain properties would be discovered later.
- Properties of elements repeat in a regular, or periodic, pattern.
- Moseley rearranged the elements in order of increasing atomic number.
- The periodic law states that the repeating chemical and physical properties of elements relate to and depend on elements' atomic numbers.
- Elements in the periodic table are classified as metals, nonmetals, and metalloids.
- Each element has a chemical symbol.
- A horizontal row of elements is called a *period.*
- Physical and chemical properties of elements change across each period.
- A vertical column of elements is called a *group* or *family.*
- Elements in a group usually have similar properties.

Name________________________________ Class _________________ Date _________________

Skills Worksheet

# Vocabulary and Section Summary

## Grouping the Elements

### VOCABULARY

**In your own words, write a definition of the following terms in the space provided.**

1. alkali metal

_______________________________________________

_______________________________________________

2. alkaline-earth metal

_______________________________________________

_______________________________________________

3. halogen

_______________________________________________

_______________________________________________

4. noble gas

_______________________________________________

_______________________________________________

### SECTION SUMMARY

**Read the following section summary.**

- Alkali metals (Group 1) are the most reactive metals. Atoms of the alkali metals have one electron in their outer level.
- Alkaline-earth metals (Group 2) are less reactive than the alkali metals are. Atoms of the alkaline-earth metals have two electrons in their outer level.
- Transition metals (Groups 3–12) include most of the well-known metals and the lanthanides and actinides.
- Groups 13–16 contain the metalloids and some metals and nonmetals.
- Halogens (Group 17) are very reactive nonmetals. Atoms of the halogens have seven electrons in their outer level.
- Noble gases (Group 18) are unreactive nonmetals. Atoms of the noble gases have a full set of electrons in their outer level.
- Hydrogen is set off by itself. Its properties do not match the properties of any one group.

Skills Worksheet

# Directed Reading A

## Section: Electrons and Chemical Bonding

### COMBINING ATOMS THROUGH CHEMICAL BONDING

_____ 1. Which of the following substances results from combining atoms of carbon, hydrogen, and oxygen?
a. sugar
b. water
c. salt
d. sulfuric acid

_____ 2. Which of the following is NOT true about electrons when chemical bonds form?
a. Electrons are shared.
b. Electrons are lost.
c. Electrons are destroyed
d. Electrons are gained.

_____ 3. Which of the following is an interaction that holds two atoms together?
a. chemical hold
b. chemical bond
c. chemical interaction
d. bond of chemicals

4. The joining of atoms to form new substances is called

____________________.

5. An explanation of a phenomenon that is based on observation, experimentation, and reasoning is a(n) ____________________.

6. People can use ______________________ to discuss theories of how and why atoms form bonds.

### ELECTRON NUMBER AND ORGANIZATION

_____ 7. How can you determine the number of electrons in an atom?
a. valence number
c. chemical number
b. atomic number
d. ionic number

_____ 8. How many valence electrons are in an oxygen atom?
a. 2
c. 6
b. 4
d. 8

_____ 9. What do elements within a group number have the same number of?
a. valance electrons
c. neutrons
b. protons
d. atoms

**Match the correct description with the correct term. Write the letter in the space provided.**

_____ 10. an electron in the outermost energy level

_____ 11. number of protons in an atom

_____ 12. family on the periodic table to which an element belongs

a. group
b. valence electrons
c. atomic number

13. Which electrons in an atom make chemical bonds? Why?

______________________________________________________________

______________________________________________________________

14. How can the periodic table help you determine the number of valence electrons?

______________________________________________________________

______________________________________________________________

## TO BOND OR NOT TO BOND

_____ 15. What determines whether an atom will form bonds?

a. number of electrons
b. number of valence electrons
c. number of protons
d. number of neutrons

_____ 16. Which group on the periodic table contains elements that do not normally form chemical bonds?

a. Group 2
b. Group 6
c. Group 10
d. Group 18

17. The outermost energy level of an atom is considered full if the level contains

____________________ electrons.

18. Helium atoms only need ____________________ valence electrons to have a filled outermost energy level.

19. The first energy level of any atom can only hold ____________________ electrons.

20. Why is it uncommon for noble gases to form chemical bonds?

______________________________________________________________

______________________________________________________________

21. Which is more likely to form bonds, an atom with 8 valence electrons or an atom with less than 8 valence electrons?

______________________________________________________________________

______________________________________________________________________

22. How can atoms with fewer than 8 valance electrons fill their outermost energy level? Use either sulfur or magnesium to explain the process.

______________________________________________________________________

______________________________________________________________________

______________________________________________________________________

______________________________________________________________________

______________________________________________________________________

______________________________________________________________________

Name________________________________ Class __________________ Date __________________

Skills Worksheet

# Directed Reading A

## Section: Ionic Bonds

### FORMING IONIC BONDS

1. A chemical bond that forms when electrons are transferred from one atom to another is a(n) ______________________.

2. Charged particles that form when atoms gain or lose electrons are ______________________.

3. A transfer of electrons between atoms changes the number of electrons in an atom, but the number of ______________________ stays the same.

4. Why is an atom neutral?

______________________________________________________________

______________________________________________________________

______________________________________________________________

5. Why are ions charged particles and thus no longer neutral?

______________________________________________________________

______________________________________________________________

______________________________________________________________

______________________________________________________________

### FORMING POSITIVE IONS

_____ 6. When atoms lose electrons through an ionic bond, they become
- a. positively charged.
- b. neutral.
- c. negatively charged.
- d. uncharged.

7. Most metals have few ______________________ and form positive ions.

8. If a sodium atom loses its only valence electron to another atom, the sodium atom becomes a sodium ______________________.

9. A sodium ion has a charge of ______________________.

10. The chemical symbol for a sodium ion is ______________________.

11. When electrons pull away from atoms, ______________________ is needed.

12. Where does the energy needed to take electrons from metals come from?

____________________________________________________________

**FORMING NEGATIVE IONS**

_____ 13. Some atoms gain electrons during chemical changes and have a
a. positive charge.
b. negative charge.
c. neutral charge.
d. chemical charge.

_____ 14. The symbol for oxide is $O^{2-}$. How many electrons did the oxygen atom gain?
a. 0
b. 1
c. 2
d. 3

_____ 15. What ending is used for the names of negative ions?
a. *-ion*
b. *-ade*
c. *-ide*
d. *-ite*

16. Atoms of Group ______________________ elements give off the most energy when they gain an electron.

17. When is energy given off by most nonmetals?

____________________________________________________________

____________________________________________________________

18. When does an ionic bond form between a metal and a nonmetal?

____________________________________________________________

____________________________________________________________

____________________________________________________________

**IONIC COMPOUNDS**

_____ 19. When ions bond, they form a repeating three-dimensional patterned called a(n)
a. compound.
b. crystal lattice.
c. chemical bond.
d. ionic bond.

20. Why does the compound formed by an ionic bond have a neutral charge when the ions that bond are charged?

_______________________________________________________________

_______________________________________________________________

_______________________________________________________________

21. List three properties of ionic compounds within a crystal lattice.

_______________________________________________________________

_______________________________________________________________

_______________________________________________________________

Name_______________________________ Class _________________ Date _________________

Skills Worksheet

# Directed Reading A

## Section: Covalent and Metallic Bonds

### COVALENT BONDS

_____ 1. What is formed when atoms share one or more pairs of electrons?
a. covalent bond
b. covalent compound
c. ionic bond
d. electric bond

_____ 2. What usually consists of two or more atoms joined in a definite ratio?
a. bond
b. valence electron
c. atom
d. molecule

3. A model that shows only the valence electrons in an atom is a(n)

_________________________.

### COVALENT COMPOUNDS AND MOLECULES

4. What is the relationship between diatomic molecules and diatomic elements? Name one example of a diatomic element.

_______________________________________________________________

_______________________________________________________________

_______________________________________________________________

_______________________________________________________________

5. What is the smallest particle into which covalent bonds can be divided?

_______________________________________________________________

6. Name two examples of complex molecules.

_______________________________________________________________

### METALLIC BONDS

7. A bond formed by the attraction between positively-charged metal ions and the electrons in the metal is a(n) _________________________.

8. What allows valence electrons in metals to move throughout the metal?

_______________________________________________________________

_______________________________________________________________

_______________________________________________________________

## PROPERTIES OF METALS

_____ 9. What property gives metals the ability to be drawn into wires?

a. malleability
b. conductivity
c. ductility
d. electricity

10. The property of _______________________ means that the metal can be hammered into sheets.

11. Give an example of how metallic bonding allows metals to conduct electric current.

_______________________________________________________________

_______________________________________________________________

_______________________________________________________________

_______________________________________________________________

12. Why doesn't a piece of metal break when it is bent?

_______________________________________________________________

_______________________________________________________________

_______________________________________________________________

Name________________________________Class ________________ Date ________________

Skill Worksheet

# Vocabulary and Section Summary

## Electrons and Chemical Bonding

### VOCABULARY

**In your own words, write a definition of the following terms in the space provided.**

1. chemical bonding

______________________________________________________________________

______________________________________________________________________

2. chemical bond

______________________________________________________________________

______________________________________________________________________

3. valence electron

______________________________________________________________________

______________________________________________________________________

### SECTION SUMMARY

**Read the following section summary.**

- Chemical bonding is the joining of atoms to form new substances. A chemical bond is an interaction that holds two atoms together.
- A valence electron is an electron in the outermost energy level of an atom.
- Most atoms form bonds by gaining, losing, or sharing electrons until they have 8 valence electrons. Atoms of some elements need only 2 electrons to fill their outermost level.

Name________________________________ Class __________________ Date __________________

Skill Worksheet

# Vocabulary and Section Summary

## Ionic Bonds

### VOCABULARY

**In your own words, write a definition of the following terms in the space provided.**

1. ionic bond

_______________________________________________________________

_______________________________________________________________

2. ion

_______________________________________________________________

_______________________________________________________________

3. crystal lattice

_______________________________________________________________

_______________________________________________________________

### SECTION SUMMARY

**Read the following section summary.**

- An ionic bond is a bond that forms when electrons are transferred from one atom to another. During ionic bonding, the atoms become oppositely charged ions.
- Ionic bonding usually occurs between atoms of metals and atoms of nonmetals.
- Energy is needed to remove electrons from metal atoms. Energy is released when most nonmetal atoms gain electrons.

Name________________________________Class __________________ Date __________________

Skill Worksheet

# Vocabulary and Section Summary

## Covalent and Metallic Bonds

### VOCABULARY

**In your own words, write a definition of the following terms in the space provided.**

1. covalent bond

______________________________________________________________

______________________________________________________________

2. molecule

______________________________________________________________

______________________________________________________________

3. metallic bond

______________________________________________________________

______________________________________________________________

### SECTION SUMMARY

**Read the following section summary.**

- In covalent bonding, two atoms share electrons. A covalent bond forms when atoms share one or more pairs of electrons.
- Covalently bonded atoms form a particle called a molecule. A molecule is the smallest particle of a compound that has the chemical properties of the compound.
- In metallic bonding, the valence electrons move throughout the metal. A bond formed by the attraction between positive metal ions and the electrons in the metal is a metallic bond.
- Properties of metals include conductivity, ductility, and malleability.

Name______________________________ Class ________________ Date ________________

Skills Worksheet

# Directed Reading A

## Section: Forming New Substances

1. The color of leaves that contain chlorophyll is ______________________.

2. Why are leaves orange and yellow in the fall?

_______________________________________________

_______________________________________________

_______________________________________________

### CHEMICAL REACTIONS

_____ 3. Which of the following names the process by which chlorophyll breaks down into new substances?

a. chemical substance
b. chemical reaction
c. chemical mixture
d. chemical solution

4. A process in which one or more substances change to form new substances is called a(n) ______________________.

5. How do the properties of the new substances compare with the properties of the original substances after a chemical change takes place?

_______________________________________________

_______________________________________________

_______________________________________________

6. A solid substance that is formed in a solution is called a(n) ______________________.

**Match the correct example of a chemical reaction with the correct clue. Write the letter in the space provided.**

_____ 7. thermal energy produced by a fire

_____ 8. precipitate

_____ 9. bubbles

_____ 10. white spots caused by bleach

a. color change
b. energy change
c. solid formation
d. gas formation

11. What can you conclude is happening if a reaction has more than one of the signs mentioned above?

_______________________________________________________________

_______________________________________________________________

_______________________________________________________________

12. What is the most important sign that a chemical reaction is occurring?

_______________________________________________________________

_______________________________________________________________

_______________________________________________________________

13. When a gas is given off as a liquid boils, it is an example of a

_____________________ change, not a _____________________ reaction.

**BONDS: HOLDING MOLECULES TOGETHER**

14. What is a chemical bond?

_______________________________________________________________

_______________________________________________________________

_______________________________________________________________

15. What is the relationship between a chemical reaction and the making and breaking of chemical bonds?

_______________________________________________________________

_______________________________________________________________

_______________________________________________________________

_______________________________________________________________

16. What makes chemical bonds break?

_______________________________________________________________

_______________________________________________________________

_______________________________________________________________

_______________________________________________________________

17. How many atoms make up a diatomic molecule?

_______________________________________________________________

Name______________________________ Class ________________ Date ________________

Skills Worksheet

# Directed Reading A

## Section: Chemical Formulas and Equations

### CHEMICAL FORMULAS

_____ 1. About how many elements make up all known substances?
a. 100
b. 80
c. 60
d. 50

_____ 2. The subscript in the chemical formula $H_2O$ tells you there are two
a. atoms of hydrogen in the molecule.
b. electrons on the hydrogen atom in the molecule.
c. elements in the molecule.
d. atoms of oxygen in the molecule.

_____ 3. What is the chemical formula for oxygen?
a. $O_2$
b. $C_6H_{12}O_6$
c. $H_2O$
d. $Ca(NO_3)_2$

_____ 4. What is the chemical formula for water?
a. $O_2$
b. $C_6H_{12}O_6$
c. $H_2O$
d. $Ca(NO_3)_2$

_____ 5. What is the chemical formula for glucose?
a. $O_2$
b. $C_6H_{12}O_6$
c. $H_2O$
d. $Ca(NO_3)_2$

6. A combination of chemical symbols and numbers that represent a substance is called a(n) ________________________.

7. What does a chemical formula show?

______________________________________________________________________

______________________________________________________________________

______________________________________________________________________

8. Covalent compounds are usually composed of two ________________________.

9. The formula for dinitrogen monoxide is ________________________.

10. The formula for carbon dioxide is ________________________.

11. Ionic compounds are composed of a(n)________________________ and a(n) ________________________.

12. The overall charge of an ionic compound is ________________________.

**Write the formula for each of the following ionic compounds.**

13. sodium chloride ______________________________________________

14. magnesium chloride ___________________________________________

## CHEMICAL EQUATIONS

15. What do musical notations and chemical equations have in common?

______________________________________________________________

______________________________________________________________

______________________________________________________________

______________________________________________________________

16. When chemical symbols and formulas are used as a shortcut to describe a chemical reaction, it is called a(n) ______________________.

17. A substance that forms in a chemical reaction is called a(n) ______________________.

18. A substance or molecule that participates in a chemical reaction is called a(n) ______________________.

19. When carbon reacts with oxygen to form carbon dioxide, carbon dioxide is the ______________________.

20. What will happen if the wrong chemical symbol or formula is used in a chemical equation?

______________________________________________________________

______________________________________________________________

______________________________________________________________

______________________________________________________________

21. In a chemical reaction, ______________________ are never gained or lost.

22. Antoine Lavoisier's work led to the ______________________.

23. The number placed in front of a chemical symbol or formula is called a(n) ______________________.

24. Chemical equations must be balanced. Why?

_______________________________________________

_______________________________________________

_______________________________________________

_______________________________________________

_______________________________________________

25. How many oxygen atoms are contained in the formula $2CO_2$?

_______________________________________________

Name______________________________Class _________________ Date _________________

Skills Worksheet

# Directed Reading A

## Section: Types of Chemical Reactions

_____ 1. Which of the following is NOT a type of chemical reaction?
a. synthesis
b. decomposition
c. single-displacement
d. double-decomposition

2. What do all four types of reactions have in common?

______________________________________________________________

______________________________________________________________

______________________________________________________________

______________________________________________________________

### SYNTHESIS REACTIONS

3. When two or more substances combine to form one new compound, it is called a(n) ______________________.

### DECOMPOSITION REACTIONS

4. When a single compound breaks down to form two or more simpler substances, it is called a(n) ______________________.

### SINGLE-DISPLACEMENT REACTIONS

5. When an element replaces another element in a compound, it is called a(n) ______________________.

6. How is a person who cuts in on a dancing couple like a single-replacement reaction?

______________________________________________________________

______________________________________________________________

______________________________________________________________

7. In a single-displacement reaction, a(n)______________________ reactive element can replace a(n) ______________________ reactive element from a compound.

8. Most single-displacement reactions involve ______________________.

## DOUBLE-DISPLACEMENT REACTIONS

9. In a double-displacement reaction, a(n) ______________________, or a(n) ______________________ forms from the exchange of ions between two compounds.

**Match the correct description with the correct term. Write the letter in the space provided.**

_____ 10. Zinc reacts with hydrochloric acid to form zinc chloride and hydrogen.

_____ 11. Carbonic acid decomposes to form water and carbon dioxide.

_____ 12. Sodium reacts with chlorine to form sodium chloride.

_____ 13. Sodium fluoride and silver chloride are formed from the reaction of sodium chloride with silver fluoride.

a. decomposition
b. double-displacement
c. single-displacement
d. synthesis

Name________________________________ Class __________________ Date __________________

Skills Worksheet

# Directed Reading A

## Section: Energy and Rates of Chemical Reactions

1. All chemical reactions either give off or absorb ______________________.

### REACTIONS AND ENERGY

2. Why is chemical energy a part of all chemical reactions?

_______________________________________________________________

_______________________________________________________________

3. When energy is released during a chemical reaction, it is called a(n)

______________________ reaction.

4. Give one example of the types of energy released in exothermic reactions.

_______________________________________________________________

_______________________________________________________________

5. When energy is taken in during a chemical reaction, it is called

a(n)______________________ reaction.

6. Photosynthesis is an example of a(n) ______________________ process.

7. What does the law of conservation of energy state?

_______________________________________________________________

_______________________________________________________________

_______________________________________________________________

8. If energy can be neither created nor destroyed in a chemical reaction, what can happen to the energy?

_______________________________________________________________

_______________________________________________________________

_______________________________________________________________

9. What happens to the energy taken in during endothermic reactions?

_______________________________________________________________

_______________________________________________________________

## RATES OF REACTIONS

10. The speed at which new particles form is called the

______________________.

11. The smallest amount of energy needed to start a chemical reaction is called

______________________.

12. Name one source of activation energy.

______________________________________________________________

## FACTORS AFFECTING RATES OF REACTIONS

13. What four factors affect how rapidly a chemical reaction takes place?

______________________________________________________________

______________________________________________________________

______________________________________________________________

14. As temperature increases, the rate of reaction ______________________.

15. A measure of the amount of one substance that is dissolved in another is called

______________________.

16. How does increasing concentration increase the rate of reaction?

______________________________________________________________

______________________________________________________________

______________________________________________________________

17. The amount of exposed surface of a substance is called

______________________.

18. How can you increase the surface area of a solid reactant?

______________________________________________________________

______________________________________________________________

______________________________________________________________

19. A substance that slows down or stops a chemical reaction is called a(n)

______________________.

20. Give one example of an inhibitor.

______________________________________________________________

21. A substance that speeds up a reaction without being permanently changed is called a(n) ______________________.

22. How can the rate of a chemical reaction be increased?

_______________________________________________________________

_______________________________________________________________

_______________________________________________________________

Name_______________________________ Class _________________ Date _________________

Skills Worksheet

# Vocabulary and Section Summary

## Forming New Substances

### VOCABULARY

**In your own words, write a definition of the following terms in the space provided.**

1. chemical reaction

_______________________________________________________________

_______________________________________________________________

2. precipitate

_______________________________________________________________

_______________________________________________________________

### SECTION SUMMARY

**Read the following section summary.**

- A chemical reaction is a process by which substances change to produce new substances with new chemical and physical properties.
- Signs that indicate a chemical reaction has taken place are a color change, formation of a gas or a solid, and production of energy.
- During a reaction, bonds are broken, atoms are rearranged, and new bonds are formed.

Name________________________________Class __________________ Date ________________

Skills Worksheet

# Vocabulary and Section Summary

## Chemical Formulas and Equations

### VOCABULARY

**In your own words, write a definition of the following terms in the space provided.**

1. chemical formula

_______________________________________________________________

_______________________________________________________________

2. chemical equation

_______________________________________________________________

_______________________________________________________________

3. reactant

_______________________________________________________________

_______________________________________________________________

4. product

_______________________________________________________________

_______________________________________________________________

5. law of conservation of mass

_______________________________________________________________

_______________________________________________________________

### SECTION SUMMARY

**Read the following section summary.**

- A chemical formula uses symbols and subscripts to describe the makeup of a compound.
- Chemical formulas can often be written from the names of covalent and ionic compounds.
- A chemical equation uses chemical formulas, chemical symbols, and coefficients to describe a reaction.
- Balancing an equation requires that the same numbers and kinds of atoms be on each side of the equation.
- A balanced equation illustrates the law of conservation of mass: mass is neither created nor destroyed during ordinary physical and chemical changes.

Name________________________________Class ___________________ Date ___________________

Skills Worksheet

# Vocabulary and Section Summary

## Types of Chemical Reactions

### VOCABULARY

**In your own words, write a definition of the following terms in the space provided.**

1. synthesis reaction

_______________________________________________________________________________

_______________________________________________________________________________

2. decomposition reaction

_______________________________________________________________________________

_______________________________________________________________________________

3. single-displacement reaction

_______________________________________________________________________________

_______________________________________________________________________________

4. double-displacement reaction

_______________________________________________________________________________

_______________________________________________________________________________

### SECTION SUMMARY

**Read the following section summary.**

- A synthesis reaction is a reaction in which two or more substances combine to form a compound.
- A decomposition reaction is a reaction in which a compound breaks down to form two or more simpler substances.
- A single-displacement reaction is a reaction in which an element takes the place of another element that is part of a compound.
- A double-displacement reaction is a reaction in which ions in two compounds exchange places.

Name________________________________ Class __________________ Date __________________

Skills Worksheet

# Vocabulary and Section Summary

## Energy and Rates of Chemical Reactions

### VOCABULARY

**In your own words, write a definition of the following terms in the space provided.**

1. exothermic reaction

_______________________________________________

_______________________________________________

2. endothermic reaction

_______________________________________________

_______________________________________________

3. law of conservation of energy

_______________________________________________

_______________________________________________

4. activation energy

_______________________________________________

_______________________________________________

5. inhibitor

_______________________________________________

_______________________________________________

6. catalyst

_______________________________________________

_______________________________________________

## SECTION SUMMARY

**Read the following section summary.**

- Energy is given off in exothermic reactions.
- Energy is used in an endothermic reaction.
- The law of conservation of energy states that energy is neither created nor destroyed.
- Activation energy is the energy needed for a reaction to occur.
- The rate of a chemical reaction is affected by temperature, concentration, surface area, and the presence of an inhibitor or catalyst.

Name______________________________ Class ________________ Date ________________

Skills Worksheet

# Directed Reading A

## Section: Ionic and Covalent Compounds

_____ 1. What is the force of attraction that holds atoms or ions together called?
a. valence electrons
b. ionic compounds
c. chemical bond
d. compound cement

_____ 2. What are the electrons found in the outermost energy levels of an atom called?
a. valence electrons
b. ionic electrons
c. covalent electrons
d. compound electrons

### IONIC COMPOUNDS AND THEIR PROPERTIES

_____ 3.An ionic bond is an attraction between
a. positively charged ions.
b. oppositely charged ions.
c. negatively charged ions.
d. metallic ions.

_____ 4. When a metal meets a nonmetal, electrons are transferred and the metal atoms become
a. positively charged.
b. neutral.
c. negatively charged.
d. oppositely charged.

_____ 5. When a metal meets a nonmetal, the nonmetal atom becomes
a. positively charged.
b. neutral.
c. negatively charged.
d. oppositely charged.

_____ 6. Table salt is formed when an electron is transferred from a sodium atom to a
a. metal atom.
b. chlorine atom.
c. nonmetal atom.
d. positively charged atom.

_____ 7. Ionic compounds tend to be brittle solids
a. at room temperature.
b. at high temperatures.
c. outdoors.
d. when wet.

_____ 8. In a crystal lattice each ion is bonded to the
a. pattern it is made with.
b. ions around it.
c. compound around it
d. crystal's edge.

_____ 9. When an ionic compound is hit, the pattern shifts, ions repel each other and the crystal
a. becomes more solid.
b. forms a new lattice.
c. breaks apart.
d. becomes bonded.

_____ 10. Because strong ionic bonds hold ions together, ionic compounds have
a. a low melting point.
b. a lukewarm melting point.
c. a high melting point.
d. a variable melting point.

_____ 11. Many ionic compounds dissolve easily
a. in air.
b. at high temperatures.
c. in water.
d. in electric current.

12. When an ionic compound dissolves in water, why can it conduct electric current?

_____________________________________________

_____________________________________________

## COVALENT COMPOUNDS AND THEIR PROPERTIES

_____ 13. Covalent compounds are formed when a group of atoms share
a. uncharged particles.
b. neutrons.
c. protons.
d. electrons.

_____ 14. Compared with ionic bonds, covalent bonds are
a. weaker.
b. stronger.
c. larger.
d. smaller.

_____ 15. The group of atoms that make up a covalent compound is called a(n)
a. bond.
b. electron.
c. molecule.
d. atom.

16. What does it mean if a substance is not soluble in water?

_____________________________________________

_____________________________________________

17. Why are covalent compounds often not soluble in water?

_____________________________________________

_____________________________________________

18. Why do covalent compounds have lower melting points than ionic compounds?

_____________________________________________

_____________________________________________

_____________________________________________

_____________________________________________

19. Why doesn't sugar dissolved in water conduct electric current?

_______________________________________________

_______________________________________________

_______________________________________________

_______________________________________________

20. How are acids that have been dissolved in water able to conduct an electric current?

_______________________________________________

_______________________________________________

_______________________________________________

Name____________________ Class ____________ Date ____________

Skills Worksheet

# Directed Reading A

## Section: Acids and Bases

### ACIDS AND THEIR PROPERTIES

_____ 1. What is any compound that increases the number of hydronium ($H_3O^+$) ions dissolved in water called?
   a. base
   b. acid
   c. indicator
   d. neutral

_____ 2. What does each hydrogen ion bond with to form hydronium ions?
   a. an oxygen particle
   b. a water molecule
   c. an acid
   d. tea

_____ 3. What do hydrogen ions, $H^+$, form when they bond to water molecules, $H_2O$?
   a. hydrogen ions, $H^+$
   b. hydronium ions, $H_3O^+$
   c. water molecules, $H_2O$
   d. bases

_____ 4. What flavor do acids have?
   a. sweet
   b salty
   c. sour
   d. crunchy

_____ 5. Why should a person NEVER taste or touch an unknown chemical?
   a. many are flavorless
   b. many are too sweet
   c. many are corrosive
   d. many are too salty

_____ 6. What can corrosive substances destroy?
   a. sour things
   b. poisons
   c. lemons
   d. body tissues and clothing

_____ 7. A compound that can reversibly change color depending on conditions such as pH is called a(n)
a. indicator.
b. color meter.
c. color changer.
d. water molecule.

_____ 8. Two commonly used indicators are bromthymol blue and
a.hydrochloric acid.
b. silver nitrate.
c. litmus paper.
d. color changer.

_____ 9. What color does blue litmus paper turn when acid is added to it?
a. green
b. red
c. blue
d. orange

_____ 10. What is produced when acids react with some metals?
a. oxygen gas
b. metals
c. silver crystals
d. hydrogen gas

_____ 11. Since acids form hydronium ions in water, solutions of acids can
a. make oxygen.
b. break apart water molecules.
c. conduct electric current.
d. straighten hair.

**Match the correct acid with the product it is used in. Write the letter in the space provided.**

_____ 12. rubber

_____ 13. car batteries

_____ 14. orange juice

_____ 15. swimming pools

_____ 16. soft drinks

a. sulfuric acid
b. nitric acid
c. hydrochloric acid
d. citric acid
e. carbonic acid

## BASES AND THEIR PROPERTIES

_____ 17. Any compound that increases the number of hydroxide ions when dissolved in water is a(n)
a. gas.
b. sodium.
c. acid.
d. base.

_____ 18. The properties of bases include a bitter taste and a(n)
a. strong bond.
b. slippery feel.
c. hydroxide lattice.
d. unpleasant odor.

_____ 19. What should you NEVER do to identify a chemical?
a. add salt to it
b. use an indicator
c. taste or touch
d. look in a book

_____ 20. What color does a base change red litmus paper to?
a. bluc
b. purple
c. green
d. orange

_____ 21. Because bases increase the number of hydroxide ions, OH–, solutions of bases can
a. indicate temperature.
b. split atoms.
c. conduct electric current.
d. stop electric current.

**Match the correct base with the product it is used in. Write the letter in the space provided.**

_____ 22. soap

_____ 23. antacids

_____ 24.cement

a. magnesium hydroxide
b. sodium hydroxide
c. calcium hydroxide

Name________________________________Class __________________ Date __________________

Skills Worksheet

# Directed Reading A

## Section: Solutions of Acids and Bases

### STRENGTHS OF ACIDS AND BASES

_____ 1. What is the amount of acid or base dissolved in water called?
a. concentration
b. strength
c. pH
d. neutralization

_____ 2. When an acid or base dissolves in water, what is dependent on the number of molecules that break apart?
a. its concentration
b. its weakness
c. its durability
d. its strength

_____ 3 In what kind of solution do all the molecules of an acid break apart in water?
a. a strong acid
b. a strong base
c. a weak acid
d. a weak base

_____ 4. In what kind of solution do only a few of the molecules of an acid break apart in water?
a. a strong acid
b. a strong base
c. a weak acid
d. a weak base

_____ 5. In what kind of solution do all the molecules of a base break apart?
a. a strong acid
b. a strong base
c. a weak acid
d. a weak base

_____ 6. What is a solution called when only a few molecules of a base break apart?
a. a strong acid
b. a strong base
c. a weak acid
d. a weak base

### ACIDS, BASES, AND NEUTRALIZATION

_____ 7. What is the reaction between acids and bases called?
a. neutralization reaction
b. explosion
c. strength
d. evaporation

_____ 8. What do the H+ ions of an acid and the OH-ions of a base form when they react?
a. oxygen
b. water
c. sugar
d. hydrogen gas

_____ 9. What can show whether a solution contains an acid or a base?
a. an indicator
b. pure water
c. antacids
d. salt

10. A value that is used to express the acidity or basicity (alkalinity) of a system is called ______________________.

11. The pH of a solution shows the concentration of what type of ion?

______________________________________________________________

______________________________________________________________

12. What is the pH of a neutral solution?

______________________________________________________________

______________________________________________________________

13. What type of solution has a pH greater than 7?

______________________________________________________________

______________________________________________________________

14. What type of solution has a pH less than 7?

______________________________________________________________

______________________________________________________________

15. What are three examples of common materials with a pH of less than 7?

______________________________________________________________

______________________________________________________________

______________________________________________________________

16. What are three examples of common materials with a pH of more than 7?

______________________________________________________________

______________________________________________________________

______________________________________________________________

**For each organism listed, write the preferred pH or pH range.**

______________________ 17. pine trees

______________________ 18. lettuce

______________________ 19. fish

20. How does acid rain form, and what is its effect on nature?

_______________________________________________________________

_______________________________________________________________

_______________________________________________________________

**SALTS**

21. What two substances are produced when an acid neutralizes a base?

_______________________________________________________________

_______________________________________________________________

22. What is a salt and how does it form?

_______________________________________________________________

_______________________________________________________________

23. Name two salts and tell what they are used for.

_______________________________________________________________

_______________________________________________________________

_______________________________________________________________

_______________________________________________________________

Name_______________________________ Class _______________ Date _______________

Skills Worksheet

# Directed Reading A

## Section: Organic Compounds

_____ 1. What is a covalent compound composed of carbon-based molecules called?
a. hydrogen atom
b. oxygen atom
c. organic compound
d. valence electron

### THE FOUR BONDS OF A CARBON ATOM

_____ 2. How many valence electrons does each carbon atom have?
a. three
b. two
c. six
d. four

_____ 3. What do structural formulas show about the atoms in a molecule of a compound?
a. what colors the atoms are
b. how the atoms are connected
c. how heavy the atoms are
d. what size the atoms are

_____ 4. What do the backbones of some compounds have hundreds or thousands of?
a. carbon atoms
b. carbon molecules
c. structural formulas
d. acid ions

### HYDROCARBONS AND OTHER ORGANIC COMPOUNDS

_____ 5. What is an organic compound composed only of carbon and hydrogen called?
a. molecule
b. electron
c. hydrocarbon
d. single bond

_____ 6. What is a hydrocarbon in which each carbon atom in the molecule shares a single bond with each of the four other atoms called?
a. unsaturated hydrocarbon
b. saturated hydrocarbon
c. bonded hydrocarbon
d. unbonded hydrocarbon

_____ 7. What is another name for a saturated hydrocarbon?
a. carbon atom
b. alkane
c. triple bond
d. atomic bond

_____ 8. What is a hydrocarbon in which at least one pair of carbon atoms share a double or triple bond called?
a. unsaturated hydrocarbon
b. saturated hydrocarbon
c. bonded hydrocarbon
d. unbonded hydrocarbon

_____ 9. What are compounds that contain two carbon atoms connected by a double bond called?
a. alkanes
b. double-binds
c. alkenes
d. alkynes

_____ 10. What are compounds that contain two carbon atoms connected by a triple bond called?
a. alkanes
b. triple-binds
c. alkenes
d. alkynes

11. What are aromatic compounds usually based on?

______________________________________________________________

______________________________________________________________

12. What kind of bonds do the atoms in a ring of benzene have?

______________________________________________________________

______________________________________________________________

13. What do aromatic hydrocarbons often have?

______________________________________________________________

______________________________________________________________

14. List three elements that other organic compounds might have in them.

______________________________________________________________

______________________________________________________________

## BIOCHEMICALS: THE COMPOUNDS OF LIFE

_____ 15. Carbohydrates, lipids, proteins and nucleic acids are the four categories of
a. living things.
b. unsaturated hydrocarbons.
c. organic compounds.
d. biochemicals.

_____ 16. Carbohydrates are biochemicals that are composed of one or more
a. saturated hydrocarbons.
b. sugar molecules.
c. organic compounds.
d. starch molecules.

_____ 17. Carbohydrates are used as a source of
a. fat.
b. genetic material.
c. energy.
d. structure.

_____ 18. Simple carbohydrates are made up of
a. simple sugars.
b. cellulose.
c. proteins.
d. lipids.

_____ 19. Complex carbohydrates are made of hundreds or thousands of
- a. lipids.
- b. sugar molecules.
- c. proteins.
- d. nucleic acids.

_____ 20. Lipids are biochemicals that do not
- a. store excess energy.
- b. make up cell membranes.
- c. dissolve in water.
- d. store vitamins.

_____ 21. Proteins are biochemicals made up of "building blocks" called
- a. sugars.
- b. amino acids.
- c. nucleic acids.
- d. lipids.

_____ 22. If a single amino acid is missing or out of place, the protein
- a. may not include sulfur.
- b. may not provide support.
- c. may not transport materials.
- d. may not function correctly.

23. List three roles that proteins have in your body and in other living things.

_______________________________________________________________

_______________________________________________________________

24. What are the largest molecules made by living organisms called?

_______________________________________________________________

_______________________________________________________________

25. What are nucleic acids made up of?

_______________________________________________________________

_______________________________________________________________

26. What is the only reason living things differ from each other?

_______________________________________________________________

_______________________________________________________________

27. Since nucleic acids contain all the information needed for a cell to make its proteins, what are nucleic acids sometimes called?

_______________________________________________________________

_______________________________________________________________

28. What are the two kinds of nucleic acids, and what are their functions?

_______________________________________________________________

_______________________________________________________________

Name________________________________Class ___________________ Date ___________________

Skills Worksheet

# Vocabulary and Section Summary

## Ionic and Covalent Compounds

### VOCABULARY

**In your own words, write a definition of the following terms in the space provided.**

1. chemical bond

_______________________________________________________________

_______________________________________________________________

2. ionic compound

_______________________________________________________________

_______________________________________________________________

3. covalent compound

_______________________________________________________________

_______________________________________________________________

### SECTION SUMMARY

**Read the following section summary.**

- Ionic compounds have ionic bonds between ions of opposite charges.
- Ionic compounds are usually brittle, have high melting points, dissolve in water, and often conduct an electric current.
- Covalent compounds have covalent bonds and consist of particles called *molecules.*
- Covalent compounds have low melting points, don't dissolve easily in water, and do not conduct electric current.

Skills Worksheet

# Vocabulary and Section Summary

## Acids and Bases

### VOCABULARY

**In your own words, write a definition of the following terms in the space provided.**

1. acid

_______________________________________________

_______________________________________________

2. indicator

_______________________________________________

_______________________________________________

3. base

_______________________________________________

_______________________________________________

### SECTION SUMMARY

**Read the following section summary.**

- An acid is a compound that increases the number of hydronium ions in solution.
- Acids taste sour, turn blue litmus paper red, react with metals to produce hydrogen gas, and may conduct an electric current when in solution.
- Acids are used for industrial purposes and in household products.
- A base is a compound that increases the number of hydroxide ions in solution.
- Bases taste bitter, feel slippery, and turn red litmus paper blue. Most solutions of bases conduct an electric current.
- Bases are used in cleaning products and acid neutralizers.

Name______________________________ Class ________________ Date ________________

Skills Worksheet

# Vocabulary and Section Summary

## Solutions of Acids and Bases

### VOCABULARY

**In your own words, write a definition of the following terms in the space provided.**

1. neutralization reaction

______________________________________________________________________

______________________________________________________________________

2. pH

______________________________________________________________________

______________________________________________________________________

3. salt

______________________________________________________________________

______________________________________________________________________

### SECTION SUMMARY

**READ THE FOLLOWING SECTION SUMMARY.**

- Every molecule of a strong acid or base breaks apart to form ions. Few molecules of weak acids and bases break apart to form ions.
- An acid and a base can neutralize one another to make salt and water.
- pH is a measure of hydronium ion concentration in a solution.
- A salt is an ionic compound formed in a neutralization reaction. Salts have many industrial and household uses.

Name______________________________ Class ________________ Date ________________

Skills Worksheet

# Vocabulary and Section Summary

## Organic Compounds

### VOCABULARY

**In your own words, write a definition of the following terms in the space provided.**

1. organic compound

______________________________________________________________________

______________________________________________________________________

2. hydrocarbon

______________________________________________________________________

______________________________________________________________________

3. carbohydrate

______________________________________________________________________

______________________________________________________________________

4. lipid

______________________________________________________________________

______________________________________________________________________

5. protein

______________________________________________________________________

______________________________________________________________________

6. nucleic acid

______________________________________________________________________

______________________________________________________________________

## SECTION SUMMARY

**Read the following section summary.**

- Organic compounds contain carbon, which can form four bonds.
- Hydrocarbons are composed of only carbon and hydrogen.
- Hydrocarbons may be saturated, unsaturated, or aromatic hydrocarbons.
- Carbohydrates are made of simple sugars.
- Lipids store energy and make up cell membranes.
- Proteins are composed of amino acids.
- Nucleic acids store genetic information and help cells make protein.

Name______________________________ Class ________________ Date ________________

Skills Worksheet

# Directed Reading A

## Section: A Solar System Is Born

1. The planets, the sun, and many moons and small bodies are part of our

____________________.

### THE SOLAR NEBULA

_____ 2. Nebulas are found in the regions of space
- a. outside the solar system.
- b. outside the force of gravity.
- c. inside stars.
- d. between stars.

_____ 3. Nebulas are mixtures of gases and
- a. water.
- b. vapor.
- c. dust.
- d. rock.

_____ 4. Which elements are mainly found in the gases of nebulas?
- a. hydrogen and helium
- b. hydrogen and oxygen
- c. carbon dioxide and helium
- d. carbon dioxide and oxygen

5. The matter of a nebula is held together by the force of

____________________.

6. A measure of the average kinetic energy, or energy of motion, of the particles

in an object is ____________________ .

7. How do gravity and pressure keep a nebula from collapsing?

______________________________________________________________

______________________________________________________________

______________________________________________________________

### UPSETTING THE BALANCE

8. What two events can upset the balance between gravity and pressure in a nebula?

______________________________________________________________

9. When a nebula collapses, small regions in the cloud called

____________________are pushed together.

10. The cloud of gas and dust that formed our solar system is called the

____________________.

## HOW THE SOLAR SYSTEM FORMED

_____ 11. As the solar nebula collapsed, the attraction between its particles
a. decreased.
b. increased.
c. stayed the same.
d. reversed.

_____ 12. The center of the collapsed cloud of gas and dust became
a. very light and cool.
b. very light and hot.
c. very dense and cool.
d. very dense and hot.

_____ 13. What happened to the solar nebula over time?
a. It became cooler and lighter.
b. It stopped rotating.
c. It flattened into a rotating disk.
d. It expanded into a large sphere.

14. In the solar nebula, bits of ______________________ collided and stuck together to form small bodies.

15. The largest of the colliding bodies in the solar system are called ______________________, or small bodies.

16. Some of the largest planetesimals were far enough from the solar nebula to attract ______________________.

17. Which planets are gas giants?

_______________________________________________

_______________________________________________

18. The inner planets of our solar system are made mostly of ______________________ material.

19. Which planets are inner planets?

_______________________________________________

_______________________________________________

20. After the planets formed, the center mass of the solar nebula became so dense and hot that it formed ______________________.

21. What happened when the gas in the nebula's center stopped collapsing?

_______________________________________________

_______________________________________________

**Number the following events in the order in which they happened. Use the numbers 1–6. Write 1 for the first event that happened. Write 6 for the last event.**

_____ 22. The largest planetesimals grew in size and attracted more gas and dust.

_____ 23. The solar nebula began to collapse.

_____ 24. The sun was born. The remaining gas and dust were removed from the solar system.

_____ 25. The solar nebula rotates and flattens. It grew warmer near its center.

_____ 26. Planets began to grow as planetesimals collided with one another.

_____ 27. Planetesimals began to form.

Name________________________________ Class __________________ Date __________________

Skills Worksheet

# Directed Reading A

## Section: The Sun: Our Very Own Star

_____ 1. The sun is a large ball of gas made mostly of
 a. oxygen and carbon.
 b. hydrogen and helium.
 c. nitrogen and sulfur.
 d. carbon dioxide and oxygen.

### THE STRUCTURE OF THE SUN

**Match the correct description with the correct term. Write the letter in the space provided.**

_____ 2. the sun's outer atmosphere

_____ 3. the thin region below the sun's corona

_____ 4. the part of the sun that can be seen from Earth

_____ 5. the region of the sun where gases circulate

_____ 6. the center of the sun

_____ 7. a very dense region of the sun

a. chromosphere
b. core
c. radiative zone
d. convective zone
e. corona
f. photosphere

### ENERGY PRODUCTION IN THE SUN

_____ 8. Early scientists thought that the sun produced its energy by
 a. rotating.
 b. expanding.
 c. collapsing inward.
 d. burning fuel.

_____ 9. Scientists later thought that energy to heat the sun was released from
 a. gravity.
 b. pressure.
 c. globules.
 d. kinetic energy.

_____ 10. Albert Einstein showed that matter and energy are
 a. the same.
 b. opposites.
 c. unchanging.
 d. interchangeable.

_____ 11. What formula did Einstein use to show the relationship between matter and energy?
 a. $E = mc$
 b. $E = mc^2$
 c. $M = ec$
 d. $M = ec^2$

_____ 12. Einstein's formula states that energy equals mass times the
a. speed of light.
b. square of the speed of light.
c. fusion of hydrogen.
d. fusion of helium.

13. The process by which two or more low-mass nuclei combine to form another nucleus is called ________________________.

14. What happens as four hydrogen nuclei fuse?

______________________________________________________________________

15. Energy is produced in the center, or ________________________, of the sun.

16. Energy passes from the sun's core through a dense region called the ________________________.

17. Hot gases are carried to the sun's visible surface from a region called the ________________________.

18. Energy leaves the sun as light from a region called the ________________________.

**SOLAR ACTIVITY**

_____ 19. The circulation of gases in the sun combines with the sun's rotation to create
a. heat.
b. radiation.
c. electric fields.
d. magnetic fields.

20. Why do some areas of the photosphere become cooler than surrounding areas?

______________________________________________________________________

______________________________________________________________________

______________________________________________________________________

______________________________________________________________________

21. Cooler, dark areas of the photosphere of the sun are called ________________________.

22. The sunspot cycle lasts about ________________________ years.

23. How might sunspot activity affect Earth?

_______________________________________________________________________________

_______________________________________________________________________________

_______________________________________________________________________________

24. Regions of very high temperature and brightness on the sun's surface are

    called ______________________.

25. How might Earth be affected by the eruption of solar flares?

_______________________________________________________________________________

_______________________________________________________________________________

_______________________________________________________________________________

Name______________________________Class _______________ Date _______________

Skills Worksheet

# Directed Reading A

## Section: The Earth Takes Shape

_____ 1. What is Earth's position in the solar system?
a. It is the closest planet to the sun.
b. It is the farthest planet from the sun.
c. It is the third planet from the sun.
d. It is the seventh planet from the sun.

_____ 2. Earth is made mostly of
a. rock.
b. gases.
c. marsh.
d. volcanoes.

_____ 3. How much of Earth's surface is covered with water?
a. one-half
b. one-tenth
c. three-fourths
d. two-thirds

### FORMATION OF THE SOLID EARTH

_____ 4. What bodies in the solar system combined to form Earth?
a. nebulas
b. stars
c. asteroids
d. planetesimals

_____ 5. What happened to Earth as gravity crushed the rock at its center?
a. It formed an irregular shape.
b. It became round.
c. The moon formed.
d. It rotated faster.

6. What two things helped Earth become warmer as it formed?

______________________________________________________________

______________________________________________________________

7. What caused the rocky material inside Earth to melt?

______________________________________________________________

______________________________________________________________

## HOW THE EARTH'S LAYERS FORMED

8. As Earth's rocks melted, denser materials sank to Earth's center to form the

   ________________________.

9. Less dense materials floated to Earth's surface to form the

   ________________________.

10. The layer of earth beneath the crust is called the ________________________.

**Match the correct description with the correct term. Write the letter in the space provided.**

_____ 11. This is made up of materials such as magnesium and iron.

_____ 12. This is made up of nickel and iron.

_____ 13. This is made up of low density materials such as oxygen, silicon, and aluminum.

a. core
b. mantle
c. crust

## FORMATION OF THE EARTH'S ATMOSPHERE

_____ 14. Most of Earth's atmosphere today is mostly made up of
a. carbon dioxide and helium.
b. nitrogen and helium.
c. carbon dioxide and oxygen.
d. nitrogen and oxygen.

_____ 15. Scientists believe that Earth's early atmosphere was a mixture of carbon dioxide and
a. water vapor.
b. helium.
c. oxygen.
d. dust particles.

_____ 16. As Earth's early atmosphere changed, it probably formed from
a. meteoroids.
b. hot springs.
c. crystal rock.
d. volcanic gases.

_____ 17. As Earth's atmosphere changed, part of it may have come from icy planetesimals called
a. moons.
b. comets.
c. ozones.
d. glaciers.

18. Describe two factors that may have contributed to the formation of Earth's first oceans.

________________________________________________________________

________________________________________________________________

## THE ROLE OF LIFE

19. Today, we are protected from the sun's ultraviolet rays by a layer of Earth's atmosphere called the ______________________.

20. Earth's early atmosphere probably did not have ozone, so many ______________________ in the air and at the Earth's surface were broken apart.

21. The first ______________________ did not need oxygen and were protected by Earth's waters.

22. The process of absorbing energy from the sun and carbon dioxide from the air is called ______________________.

23. Early organisms released ______________________ during the process of making food.

24. As oxygen was added to the atmosphere, what gas was removed?

________________________________________________________________

25. When did simple plants move onto land?

________________________________________________________________

## FORMATION OF OCEANS AND CONTINENTS

_____ 26. What factor probably caused the oceans to form?
- a. Materials in Earth's crust and mantle melted.
- b. Carbon dioxide in the atmosphere condensed into water.
- c. Oxygen from early organisms condensed into water.
- d. Earth cooled enough for rain to fall.

_____ 27. How long ago did a global ocean cover the planet?
- a. 4 billion years ago
- b. 40 million years ago
- c. 10 billion years ago
- d. 100 million years ago

**Number the following events in the order in which they happened. Use the numbers 1–4. Write 1 for the first event that happened. Write 4 for the last event.**

_____ 28. The earliest continents formed as lighter rocks rose to the surface of the earth.

_____ 29. The upper mantle cooled and became denser and heavier.

_____ 30. Thermal energy in the mantle caused continents to move.

_____ 31. Continents thickened and rose above the surface of the ocean.

Name_______________________________ Class _________________ Date _________________

Skills Worksheet

# Directed Reading A

## Section: Planetary Motion

### A REVOLUTION IN ASTRONOMY

1. How does Earth's rotation determine whether it is day or night?

______________________________________________________________

______________________________________________________________

______________________________________________________________

**Match the correct definition with the correct term. Write the letter in the space provided.**

_____ 2. the spinning of a body on its axis

_____ 3. the path a body follows as it travels around another body in space

_____ 4. a complete trip along an orbit

_____ 5. the time it takes a planet to complete a single trip around the sun

a. revolution
b. rotation
c. period of revolution
d. orbit

6. According to Kepler's first law of motion, planets move in a(n)

_______________________ around the sun.

7. The maximum length of an ellipse, or a planet's orbit, is called its

_______________________.

8. A planet's maximum distance from the sun is the _______________________ of its orbit.

9. According to Kepler's second law of motion, how does a planet's distance from the sun affect its motion?

______________________________________________________________

______________________________________________________________

______________________________________________________________

10. According to Kepler's third law of motion, what information can be used to find a planet's distance from the sun?

_______________________________________________________________

_______________________________________________________________

**NEWTON TO THE RESCUE!**

_____ 11. What causes the planets that are closer to the sun to move faster?
   a. gravity
   b. heat
   c. magnetic fields
   d. solar energy

_____ 12. Newton discovered that the force of gravity depends on the distance between objects and the objects'
   a. volume.
   b. circumference.
   c. shape.
   d. mass.

_____ 13. The force of gravity between two objects increases if
   a. they have smaller masses and are farther apart.
   b. they have smaller masses and are closer together.
   c. they have larger masses and are farther apart.
   d. they have larger masses and are closer together.

_____ 14. An object's resistance in speed or direction is called
   a. mass.
   b. pressure.
   c. inertia.
   d. energy.

_____ 15. Gravity causes bodies in the solar system to
   a. repel one another.
   b. fall in a straight path.
   c. stay in orbit.
   d. move in a circular path.

Name_______________________________ Class _______________ Date _______________

Skills Worksheet

# Vocabulary and Section Summary

## A Solar System Is Born

### VOCABULARY

**In your own words, write a definition of the following terms in the space provided.**

1. nebula

_______________________________________________________________

_______________________________________________________________

2. solar nebula

_______________________________________________________________

_______________________________________________________________

### SECTION SUMMARY

**Read the following section summary.**

- The solar system formed out of a vast cloud of gas and dust called the *nebula.*
- Gravity and pressure were balanced until something upset the balance. Then the nebula began to collapse.
- Collapse of the solar nebula caused heating at the center, while planetesimals formed in surrounding space.
- The central mass of the nebula became the sun. Planets formed from the surrounding materials.

Name________________________________ Class __________________ Date __________________

Skills Worksheet

# Vocabulary and Section Summary

## The Sun: Our Very Own Star

### VOCABULARY

**In your own words, write a definition of the following terms in the space provided.**

1. nuclear fusion

_______________________________________________________________

_______________________________________________________________

2. sunspot

_______________________________________________________________

_______________________________________________________________

### SECTION SUMMARY

**Read the following section summary.**

- The sun is a large ball of gas made mostly of hydrogen and helium. The sun consists of many layers.
- The sun's energy comes from nuclear fusion that takes place in the center of the sun.
- The visible surface of the sun, or the photosphere, is very active.
- Sunspots and solar flares are the result of the sun's magnetic fields that reach space.
- Sunspot activity may affect Earth's climate, and solar flares can interact with Earth's atmosphere.

Name________________________________ Class _________________ Date _________________

Skills Worksheet

# Vocabulary and Section Summary

## The Earth Takes Shape

### VOCABULARY

**In your own words, write a definition of the following terms in the space provided.**

1. crust

_______________________________________________

_______________________________________________

2. mantle

_______________________________________________

_______________________________________________

3. core

_______________________________________________

_______________________________________________

### SECTION SUMMARY

**Read the following section summary.**

- The effects of gravity and heat created the shape and structure of Earth.
- The Earth is divided into three main layers based on composition: the crust, mantle, and core.
- The presence of life dramatically changed Earth's atmosphere by adding free oxygen.
- Earth's oceans formed shortly after the Earth did, when it had cooled off enough for rain to fall. Continents formed when lighter materials gathered on the surface and rose above sea level.

Name________________________________Class __________________ Date __________________

Skills Worksheet

# Vocabulary and Section Summary

## Planetary Motion

### VOCABULARY

**In your own words, write a definition of the following terms in the space provided.**

1. rotation

_____________________________________________________________

_____________________________________________________________

2. orbit

_____________________________________________________________

_____________________________________________________________

3. revolution

_____________________________________________________________

_____________________________________________________________

### SECTION SUMMARY

**Read the following section summary.**

- Rotation is the spinning of a planet on its axis, and revolution is one complete trip along an orbit.
- Planets move in an ellipse around the sun. The closer they are to the sun, the faster they move. The period of a planet's revolution depends on the planet's semimajor axis.
- Gravitational attraction decreases as distance increases and as mass decreases.

Name________________________________ Class __________________ Date __________________

Skills Worksheet

# Directed Reading A

## Section: Magnets and Magnetism

### PROPERTIES OF MAGNETS

1. Any material that attracts iron is a(n) ________________________.

2. The points on a magnet that have opposite magnetic qualities are the

   ________________________.

3. The magnetic pole that points to the north is the magnet's

   ________________________.

4. The magnetic pole that points to the south is the magnet's

   ________________________.

5. The force that can either push magnets apart or pull them together is

   ________________________.

6. The region around a magnet in which magnetic forces act is the

   ________________________.

**For each description below, identify the correct magnetic property. Write either *magnetic forces* or *magnetic fields* in the space provided.**

________________________ 7. come from spinning electric charges in the magnets

________________________ 8. can push magnets apart or pull them together

________________________ 9. depend on how two magnets' poles line up

________________________ 10. are regions around magnets in which magnetic forces can act

________________________ 11. shape that can be shown with lines that surround magnets

________________________ 12. are strongest at magnetic poles, where lines are closest together

## THE CAUSE OF MAGNETISM

_____ 13. Whether a material is magnetic depends on its
- a. density.
- b. atoms.
- c. shape.
- d. mass.

_____ 14. As an electron moves, it makes, or induces a(n)
- a. aurora.
- b. ferromagnet.
- c. electromagnet.
- d. magnetic field.

_____ 15. Materials in which the atoms' magnetic fields cancel each other out are
- a. aligned in domains.
- b. like iron, nickel, and cobalt.
- c. not magnetic.
- d. magnetic.

_____ 16. Which of these is true when the poles of atoms line up?
- a. The atoms cancel each other out.
- b. The atoms are arranged in a domain.
- c. The atoms make a weak magnetic field.
- d. The atoms do not become magnetic.

17. Name one thing that causes domains of a magnet's atoms to lose alignment.

______________________________________________________________

______________________________________________________________

18. How do you magnetize something made of iron, cobalt, or nickel?

______________________________________________________________

______________________________________________________________

______________________________________________________________

______________________________________________________________

19. Why do you end up with two magnets when you cut one magnet in half?

______________________________________________________________

______________________________________________________________

______________________________________________________________

______________________________________________________________

## KINDS OF MAGNETS

**Match the correct description with the correct term. Write the letter in the space provided.**

_____ 20. magnet with strong magnetic properties

_____ 21. magnet made by an electric current

_____ 22. magnet that loses magnetization easily

_____ 23. hard to magnetize, but stays magnetized

a. temporary
b. electromagnet
c. ferromagnet
d. permanent

## EARTH AS A MAGNET

24. Why can magnets point north?

_______________________________________________

_______________________________________________

_______________________________________________

_______________________________________________

_______________________________________________

_______________________________________________

_______________________________________________

25. If you put a compass on a bar magnet, the needle points to the south pole of the magnet. Explain why.

_______________________________________________

_______________________________________________

_______________________________________________

_______________________________________________

_______________________________________________

_______________________________________________

26. Why does a compass needle point to Earth's geographic north?

______________________________________________________________________

______________________________________________________________________

______________________________________________________________________

______________________________________________________________________

______________________________________________________________________

______________________________________________________________________

27. What makes Earth's magnetic field?

______________________________________________________________________

______________________________________________________________________

______________________________________________________________________

______________________________________________________________________

______________________________________________________________________

______________________________________________________________________

28. When charged particles from the sun hit oxygen and nitrogen atoms in the air,

a(n) ___________________________ is formed.

Name______________________________ Class _________________ Date _________________

Skills Worksheet

# Directed Reading A

## Section: Magnetism from Electricity

_____ 1. What kind of train uses an electromagnet to float above the track?
a. magnetic
b. maglev
c. electric
d. electronic

### THE DISCOVERY OF ELECTROMAGNETISM

_____ 2. The interaction between electricity and magnetism is called
a. electromagnetism.
d. electronic.
c. electric.
b. maglev

_____ 3. Oersted discovered that electric current produces a(n)
a. electromagnetism
b. maglev.
c. electric.
d. electronic.

_____ 4. The direction of a magnetic field produced by an electric current depends on the direction of the
a. current.
b. magnetism.
c. wire.
d. batteries.

5. Who were the two scientists who did the first research into the interaction between electricity and magnetism?

_______________________________________________________________

_______________________________________________________________

### USING ELECTROMAGNETISM

_____ 6. What are two devices that strengthen the magnctic ficld of a current-carrying wire?
a. magnetic field and magnetic force
b. solenoid and electromagnet
c. electromagnet and current
d. solenoid and current

_____ 7. A coil of wire that produces a magnetic field when carrying an electric current is called a(n)
a. electromagnet.
b. maglev.
c. solenoid.
d. magnetic field.

_____ 8. What happens to the magnetic field if more loops per meter are added to a solenoid?
a. The magnetic field becomes weaker.
b. The magnetic field becomes stronger.
c. The magnetic field turns on and off.
d. There is no change in the magnetic field.

_____ 9. A solenoid wrapped around a soft iron core is called a(n)
a. electromagnet.
b. maglev.
c. magnetic core.
d. magnetic field.

_____ 10. What happens to an electromagnet if the electric current in the solenoid wire is increased?
a. The electromagnet becomes weaker.
b. The electromagnet becomes stronger.
c. The electromagnet turns on and off.
d. There is no change in the electromagnet.

## APPLICATIONS OF ELECTROMAGNETISM

_____ 11. What is one thing that uses an electromagnet?
a. bicycle
b. doorbell
c. computer
d. solenoid

_____ 12. An electric motor changes electrical energy into what kind of energy?
a. electromagnetic
b. electronic
c. mechanical
d. magnetic

13. Explain what happens to an electromagnet when there is no current in the wire.

_______________________________________________

_______________________________________________

_______________________________________________

_______________________________________________

**Match the correct description with the correct term. Write the letter in the space provided. Some terms will not be used.**

_____ 14. a device that converts electrical energy into mechanical energy

_____ 15. attached to the armature; reverses direction of electric current

_____ 16. a loop or coil of wire that can rotate

_____ 17. used to measure current

a. galvanometer
b. armature
c. electric motor
d. commutator
e. voltmeter

Name__________ Class __________ Date __________

Skills Worksheet

# Directed Reading A

## Section: Electricity from Magnetism

### ELECTRIC CURRENT FROM A CHANGING MAGNETIC FIELD

1. What problem did both Joseph Henry and Michael Faraday work to solve?

______________________________________________

______________________________________________

2. The process that causes an electric current in a changing a magnetic field is called ______________________.

3. Describe what happened to the electric current in Michael Faraday's experiment when the battery was fully connected.

______________________________________________

______________________________________________

______________________________________________

______________________________________________

______________________________________________

______________________________________________

______________________________________________

______________________________________________

4. Describe two ways to induce a larger electric current when you move a magnet in a coil of wires.

______________________________________________

______________________________________________

______________________________________________

______________________________________________

______________________________________________

______________________________________________

______________________________________________

______________________________________________

## ELECTRIC GENERATORS

_____ 5. What device converts mechanical energy into electrical energy?
- a. electric motor
- b. electric generator
- c. electromagnetic motor
- d. magnetic motor

_____ 6. When electric current changes direction it is called a(n)
- a. generated current.
- b. electromagnetic current.
- c. alternating current.
- d. rotating current.

7. Name the four parts of a simple generator, and describe what they do.

______________________________________________________________

______________________________________________________________

______________________________________________________________

______________________________________________________________

8. Other than the size, what is one difference between power plants and electric generators?

______________________________________________________________

______________________________________________________________

______________________________________________________________

______________________________________________________________

______________________________________________________________

9. Name two sources of energy that generators convert into electrical energy.

______________________________________________________________

______________________________________________________________

**Put the following steps for generating electrical energy in order from 1 to 4. Write the appropriate numbers in the space provided.**

_____ 10. Steam turns a turbine.

_____ 11. Energy boils water into steam.

_____ 12. Electric current is induced and electrical energy is generated.

_____ 13. A turbine turns the magnet of a generator.

## TRANSFORMERS

_____ 14. A device that increases or decreases the voltage of alternating current is called a(n)

a. voltmeter.
b. generator.
c. transformer.
d. electromagnet.

15. Explain why a transformer uses different numbers of loops in its primary and secondary coils.

______________________________________________________________

______________________________________________________________

______________________________________________________________

______________________________________________________________

______________________________________________________________

16. Describe what a step-up transformer does.

______________________________________________________________

______________________________________________________________

______________________________________________________________

______________________________________________________________

______________________________________________________________

17. Describe what a step-down transformer does.

______________________________________________________________

______________________________________________________________

______________________________________________________________

______________________________________________________________

______________________________________________________________

______________________________________________________________

Name________________________________ Class __________________ Date __________________

Skills Worksheet

# Vocabulary and Section Summary

## Magnets and Magnetism

### VOCABULARY

**In your own words, write a definition of the following terms in the space provided.**

1. magnet

_______________________________________________________________

_______________________________________________________________

2. magnetic pole

_______________________________________________________________

_______________________________________________________________

3. magnetic force

_______________________________________________________________

_______________________________________________________________

### SECTION SUMMARY

**Read the following section summary.**

- All magnets have two poles. The north pole will always point to the north if allowed to rotate freely. The other pole is called the south pole.
- Like magnetic poles repel each other. Opposite magnetic poles attract.
- Every magnet is surrounded by a magnetic field. The shape of the field can be shown with magnetic field lines.
- A material is magnetic if its domains line up.
- Magnets can be classified as ferromagnets, electromagnets, temporary magnets, and permanent magnets.
- Earth acts as if it has a big bar magnet through its core. Compass needles and the north poles of magnets point to Earth's magnetic south pole, which is near Earth's geographic North Pole.
- Auroras are most commonly seen near Earth's magnetic poles because Earth's magnetic field bends inward at the poles.

Name_______________________________ Class ________________ Date ________________

Skills Worksheet

# Vocabulary and Section Summary

## Magnetism from Electricity

### VOCABULARY

**In your own words, write a definition of the following terms in the space provided.**

1. electromagnetism

______________________________________________________________

______________________________________________________________

2. solenoid

______________________________________________________________

______________________________________________________________

3. electromagnet

______________________________________________________________

______________________________________________________________

4. electric motor

______________________________________________________________

______________________________________________________________

### SECTION SUMMARY

**Read the following section summary.**

- Oersted discovered that a wire carrying a current makes a magnetic field.
- Electromagnetism is the interaction between electricity and magnetism.
- An electromagnet is a solenoid that has an iron core.
- A magnet can exert a force on a wire carrying a current.
- A doorbell, an electric motor, and a galvanometer all make use of electromagnetism.

Name______________________________ Class ________________ Date ________________

Skills Worksheet

# Vocabulary and Section Summary

## Electricity from Magnetism

### VOCABULARY

**In your own words, write a definition of the following terms in the space provided.**

1. electromagnetic induction

_______________________________________________

_______________________________________________

2. electric generator

_______________________________________________

_______________________________________________

3. transformer

_______________________________________________

_______________________________________________

### SECTION SUMMARY

**Read the following section summary.**

- Electromagnetic induction is the process of making an electric current by changing a magnetic field.
- An electric generator converts mechanical energy into electrical energy through electromagnetic induction.
- A step-up transformer increases the voltage of an alternating current. A step-down transformer decreases the voltage.
- The side of a transformer that has the greater number of loops has the higher voltage.

Name______________________________ Class ______________ Date ______________

Skills Worksheet

# Directed Reading A

## Section: Electronic Devices

### INSIDE AN ELECTRONIC DEVICE

_____ 1. What role does a circuit board play?
- a. receives information from a TV
- b. connects the parts of a circuit
- c. acts like an antenna
- d. serves as a power source

_____ 2. The LED in a remote control
- a. gives off radio waves.
- b. is only for decoration.
- c. sends information to the TV
- d. receives information from the TV

3. An LED, or ______________________, is one of the electronic components within a TV remote control.

4. A sheet of insulating material called a(n) ______________________ carries circuit elements and is inserted into electronic devices.

### SEMICONDUCTORS

5. What element or compound conducts an electric current better than an insulator does?

______________________________________________________________

______________________________________________________________

6. What happens when silicon atoms bond?

______________________________________________________________

______________________________________________________________

7. The addition of an impurity to a semiconductor is called ______________________.

8. What happens when a silicon atom is replaced with an arsenic atom? What type of semiconductor is produced?

______________________________________________________________

______________________________________________________________

9. What happens when a silicon atom is replaced with a gallium atom? What kind of semiconductor is produced?

_______________________________________________

_______________________________________________

**DIODES**

10. An electronic component that allows electric charge to move mainly in one direction is called a ________________.

11. What happens to the "extra" electrons when two layers of a diode meet?

_______________________________________________

_______________________________________________

12. How do power plants send electrical energy to homes?

_______________________________________________

_______________________________________________

13. How do diodes help change AC to DC?

_______________________________________________

_______________________________________________

_______________________________________________

_______________________________________________

**TRANSISTORS**

_____ 14. A transistor is an electronic component that
   a. decreases current.
   b. does not affect current.
   c. amplifies or increases current.
   d. blocks current.

15. A transistor can be used in many devices, including an amplifier and a(n) ________________.

16. Why is a transfer useful in an amplifier?

_______________________________________________

_______________________________________________

17. How are transistors used in devices as switches?

_______________________________________________________________

_______________________________________________________________

**INTEGRATED CIRCUITS**

_____ 18. Which of the following statements describes benefits of using integrated circuits?
a. Few circuits can fit onto one integrated circuit.
b. Devices that use integrated circuits can run at very high speeds.
c. An integrated circuit has many components on several semiconductors.
d. Electric charges moving through integrated circuits have to travel great distances.

19. An entire circuit with many components on a single semiconductor is a(n)

______________________.

20. What are some of the advantages of replacing vacuum tubes with transistors and diodes?

_______________________________________________________________

_______________________________________________________________

Name________________________________ Class __________________ Date __________________

Skills Worksheet

# Directed Reading A

## Section: Communication Technology

### COMMUNICATING WITH SIGNALS

_____ 1. One of the first electronic communication devices was the
- a. telephone.
- b. computer.
- c. telegraph
- d. typewriter

2. Something that can be used to send information is called a(n) ______________________.

3. Sometimes, a signal is sent using another signal called a(n) ______________________.

### ANALOG SIGNALS

4. A signal whose properties change without a break between values is called a(n) ______________________.

5. What is the analog signal in a telephone system?

______________________________________________________________________

6. In a telephone, the ______________________ changes the analog signal back into the sound of your voice.

7. You talk into the ______________________ of a telephone, which vibrates producing an electric current.

8. In vinyl records, the number and depth of the ______________________ in the disk represent the sound's ______________________.

9. When you play a record, the ______________________, or, ______________________ makes the electromagnet vibrate.

10. What problem occurs as a result of the stylus touching the record?

______________________________________________________________________

______________________________________________________________________

## DIGITAL SIGNALS

**Match the correct description with the correct term. Write the letter in the space provided.**

_____ 11. represented by a pulse

_____ 12. represented by a missing pulse

_____ 13. represented as a sequence of separate values

_____ 14. means "two"

_____ 15. short for binary digit

a. digital signal
b. binary
c. number 0
d. number 1
e. bit

16. Why are the pits and lands on a CD important?

_______________________________________________________________

_______________________________________________________________

17. How does a CD player work?

_______________________________________________________________

_______________________________________________________________

_______________________________________________________________

18. Why do CDs last so long?

_______________________________________________________________

_______________________________________________________________

_______________________________________________________________

## RADIO AND TELEVISION

_____ 19. In radio, a modulator
a. changes sound waves into electric current.
b. strengthens the analog signal.
c. combines the amplified analog signal with radio waves.
d. removes the radio waves from the analog signal.

_____ 20. What transmits modulated radio waves through the air?
a. microphone
b. radio tower
c. antenna
d. modulator

21. TV and radio signals can be either ______________________ or analog.

22. What role do electromagnetic waves play in radio and television broadcasts?

_______________________________________________

23. Television images are made by beams of _______________ hitting the screen.

24. What are three ways in which signals are sent to your TV?

_______________________________________________

_______________________________________________

_______________________________________________

25. Why is it better to watch digital shows on a digital display rather than on an analog display?

_______________________________________________

_______________________________________________

_______________________________________________

26. Video signals transmitted from a TV station are received by the _______________ of a TV receiver.

27. What is the role of fluorescent materials in producing an image on a color television?

_______________________________________________

_______________________________________________

28. Why are standard television sets so bulky and heavy?

_______________________________________________

_______________________________________________

_______________________________________________

29. New types of television screens, called _______________ _______________, are thinner than standard screens.

30. In a plasma display, a(n) _______________ charges thousands of cells with gases in them to produce an electric current.

31. In a plasma display, each well contains fluorescent materials that give off _______________.

Name______________________________ Class ________________ Date ________________

Skills Worksheet

# Directed Reading A

## Section: Computers

### WHAT IS A COMPUTER?

**Match the correct definition with the correct term. Write the letter in the space provided.**

_____ 1. performing an action such as adding a list of numbers

_____ 2. information you give a computer

_____ 3. final result of the work done by the computer

_____ 4. information in a computer's memory

_____ 5. electronic device that performs tasks by following instructions

a. output
b. processing
c. storage
d. input
e. computer

6. What was the first general-purpose computer called?

_______________________________________________

7. Who made the first general-purpose computer, and when was it built?

_______________________________________________

_______________________________________________

8. Why did ENIAC have to be cooled?

_______________________________________________

_______________________________________________

_______________________________________________

9. A single semiconductor chip called a(n) ______________________ controls and carries out a modern computer's instructions.

10. Why are modern computers smaller than early computers?

_______________________________________________

_______________________________________________

## COMPUTER HARDWARE

_____ 11. The parts or pieces of equipment that make up a computer are called
a. hardware.
b. software.
c. microprocessors.
d. hardwires.

**Match the correct description with the correct term. Write the letter in the space provided.**

_____ 12. permanent memory

_____ 13. device used to send information through telephone lines

_____ 14. a type of output device

_____ 15. temporary memory

_____ 16. a type of input device

_____ 17. the microprocessor in a personal computer

a. mouse
b. CPU
c. ROM
d. RAM
e. printer
f. modem

## COMPACT DISCS

18. What can you store on a CD?

_______________________________________________________________

_______________________________________________________________

_______________________________________________________________

19. Explain why CD-R and CD-RW discs work differently.

_______________________________________________________________

_______________________________________________________________

_______________________________________________________________

**Match the correct description with the correct term. Write the letter in the space provided.**

_____ 20. device used to "burn" or heat dyes on recordable compact discs

_____ 21. computer disc that can be used only once

_____ 22. computer disc that can be erased and written over again

a. CD-R
b. CD-RW
c. laser

## COMPUTER SOFTWARE

23. What is a set of instructions or commands for a computer called?

_______________________________________________

24. What are three jobs that are handled by operating-system software?

_______________________________________________

_______________________________________________

_______________________________________________

25. How are operating-system software and application software different?

_______________________________________________

_______________________________________________

## COMPUTER NETWORKS

**Match the correct description with the correct term. Write the letter in the space provided.**

_____ 26. network in which groups of computers connect to an ISP through only one line

_____ 27. huge computer network made up of millions of computers

_____ 28. part of the Internet that people know best

_____ 29. method used to find Web pages

_____ 30. part on a web page where clicking on it takes you from one page or site to another

_____ 31. device used to look at pages on the Internet

a. link
b. web browser
c. LAN
d. search engine
e. Internet
f. World Wide Web

Name______________________________ Class ________________ Date ________________

Skills Worksheet

# Vocabulary and Section Summary

## Electronic Devices

### VOCABULARY

**In your own words, write a definition of the following terms in the space provided.**

1. circuit board

______________________________________________________________________

______________________________________________________________________

2. semiconductor

______________________________________________________________________

______________________________________________________________________

3. doping

______________________________________________________________________

______________________________________________________________________

4. diode

______________________________________________________________________

______________________________________________________________________

5. transistor

______________________________________________________________________

______________________________________________________________________

6. integrated circuit

______________________________________________________________________

______________________________________________________________________

## SECTION SUMMARY

### Read the following section summary

- Circuit boards contain circuits that supply current to different parts of electronic devices.
- Semiconductors are often used in electronic devices because their conductivity can be changed by doping.
- Diodes allow current in one direction and can change AC to DC.
- Transistors are used in amplifiers and switches.
- Integrated circuits have made smaller, smarter electronic devices possible.

Name________________________________ Class ________________ Date ________________

Skills Worksheet

# Vocabulary and Section Summary

## Communication Technology

### VOCABULARY

**In your own words, write a definition of the following terms in the space provided.**

1. analog signal

_______________________________________________

_______________________________________________

2. digital signal

_______________________________________________

_______________________________________________

### SECTION SUMMARY

**Read the following section summary**

- Signals transmit information in electronic devices. Signals can be transmitted using a carrier. Signals can be analog or digital.
- Analog signals have continuous values. Telephones, record players, radios, and regular TV sets use analog signals.
- In a telephone, a transmitter changes sound waves to electric current. The current is sent across a phone line. The receiving telephone converts the signal back into a sound wave.
- Analog signals of sounds are used to make vinyl records. Changes in the groove reflect changes in the sound.
- Digital signals have discrete values, such as 0 and 1. CD players use digital signals.
- Radios and television sets use electromagnetic waves. These waves travel through the atmosphere. In a radio, the signals are converted to sound waves. In a television set, electron beams convert the signals into images on the screen.

Name______________________________Class __________________ Date ________________

Skills Worksheet

# Vocabulary and Section Summary

## Computers

### VOCABULARY

**In your own words, write a definition of the following terms in the space provided.**

1. computer

______________________________________________________________

______________________________________________________________

2. microprocessor

______________________________________________________________

______________________________________________________________

3. hardware

______________________________________________________________

______________________________________________________________

4. software

______________________________________________________________

______________________________________________________________

5. Internet

______________________________________________________________

______________________________________________________________

## SECTION SUMMARY

**Read the following section summary.**

- All computers have four basic functions: input, processing, storage, and output.
- The first general-purpose computer, ENIAC, was made of thousands of vacuum tubes and filled an entire room. Microprocessors have made it possible to have computers the size of notebooks.
- Computer hardware includes input devices, the CPU, memory, output devices, and modems.
- CD burners can store information on recordable CDs, or CD-Rs. Rewritable CDs, or CD-RWs, can be erased and reused. Both use patterns of light and dark spots.
- Computer software is a set of instructions that tell a computer what to do. The two main types are operating systems and applications. Applications include word processors, spreadsheets, and games.
- The Internet is a huge network that allows millions of computers to share information.